Microalgae

Series Editor
Françoise Gaill

Microalgae

From Future Food to Cellular Factory

Joël Fleurence

WILEY

First published 2021 in Great Britain and the United States by ISTE Ltd and John Wiley & Sons, Inc.

ISTE Ltd
27-37 St George's Road
London SW19 4EU
UK

www.iste.co.uk

John Wiley & Sons, Inc.
111 River Street
Hoboken, NJ 07030
USA

www.wiley.com

Library of Congress Control Number: 2021938404

British Library Cataloguing-in-Publication Data
A CIP record for this book is available from the British Library
ISBN 978-1-78630-587-9

Contents

Preface ix

Acknowledgments xi

Introduction xiii

Chapter 1. Biology and Ecology of Microalgae 1

1.1. Biological characteristics 1
1.1.1. General characteristics 1
1.1.2. The different groups in traditional and phylogenetic classification 3
1.1.3. The special case of cyanobacteria (Cyanophyceae) 10
1.2. Ecological features 12
1.2.1. Marine microalgae 13
1.2.2. Microalgae in brackish and freshwater environments 15
1.2.3. Microalgae in terrestrial and aerial environments 16

Chapter 2. Production Techniques 21

2.1. Production by harvesting in the natural environment 21
2.2. Production by culture in open systems 24
2.2.1. Production in open basins 24
2.2.2. Production in open raceway-type basins 25
2.2.3. Open-tank production 30
2.3. Production by culture in a closed system 31
2.3.1. Production in discontinuous mode 31
2.3.2. Production in continuous mode 34

Chapter 3. Food Valorizations 43

3.1. Animal feed. 43
3.1.1. Forage microalgae 43
3.1.2. Dietary supplements 51
3.2. Human food 56
3.2.1. Ingredients or vegetables 56
3.2.2. Dietary supplements 60
3.2.3. Functional foods 66
3.2.4. Food coloring 73
3.2.5. Regulations 74

Chapter 4. Valorized Molecules 77

4.1. Polysaccharides 77
4.2. Proteins and enzymes 83
4.2.1. Phycobiliproteins 83
4.2.2. Enzymes 87
4.3. Non-protein pigments. 89
4.4. Fat, sterols and fatty acids 90
4.5. The special case of biofuel. 94
4.5.1. Biofuel production processes 94
4.5.2. Algal species used as biosources. 99
4.5.3. The economic context 99
4.6. Other applications 101

Chapter 5. Extraction Processes 105

5.1. Conventional processes. 105
5.1.1. Ball mills. 105
5.1.2. Ultrasonication 106
5.1.3. Extraction using supercritical fluid 109
5.1.4. Extraction by microwaves. 113
5.1.5. High-pressure extraction. 114
5.1.6. Extraction facilitated by lyophilization 116
5.2. Enzymatic hydrolysis 118
5.3. Other methods 122

Chapter 6. Biotechnological Approaches 125

6.1. Biorefinery 125
6.2. Physiological forcing 127
6.3. Genetic transformation 131

Conclusion 137

References 139

Index 155

Preface

Unicellular algae and cyanobacteria are unstoppable actors of life on Earth. They are responsible for the production of half of the oxygen present on our planet. Independently of this major ecological role, microalgae are true cellular factories at the origin of the synthesis of various metabolites of interest for human activity. In particular, they produce original pigments, such as phycobiliproteins, polysaccharides, enzymes, as well as lipids and fatty acids with a long carbon chain, whose valorization as biofuels opens up a new way of valorizing these microorganisms. In addition to the biotechnological uses resulting from this cell plant concept and the resulting biorefinery, microalgae also constitute a food resource for human and animal nutrition.

The purpose of this book is to take stock of the biological and ecological characteristics of microalgae, whether marine, freshwater or even atmospheric. Cyanobacteria (e.g. *Arthrospira* sp. or *Spirulina* sp.) are also discussed, as they have long been referred to as "blue algae" under the botanical name of Cyanophyceae.

This book also assesses the production methods and current applications of microalgae and cyanobacteria, whether in the food or biotechnology fields. As far as food applications are concerned, the current uses of microalgae in animal and human nutrition in the form of food supplements are presented. The prospects for development in these application sectors are discussed in light of economic and regulatory constraints.

The biotechnological valorizations of microalgae as cell factories capable of producing molecules with high added value are also described in this book. The techniques for extracting these molecules and the new approaches for valorization, such as biorefinery, are also discussed. Finally, the biotechnological perspective of using genetically modified microalgae for the production of molecules for therapeutic purposes is developed in light of current advances in research in the field.

May 2021

Acknowledgements

I thank Yves-François Pouchus and Olivier Grovel for their illustrations.

Introduction

Cyanobacteria and microalgae are, from a logical chronological point of view, at the origin of oxygen production by photosynthetic means on our planet. Oxygen photosynthesis by cyanobacteria probably occurred 3 billion years ago (Sardet 2013). The appearance of the first eukaryotic cells, including microalgae, is estimated to have occurred 1.5–2 billion years ago. The extent of algal biological diversity in the ocean has long been unknown (Sournia *et al.* 1991). The number of microalgae species is still not definitively established, and the figures given in the scientific literature range from 70,000 to 150,000 species. Conversely, high estimates suggest a million species (De Vargas *et al.* 2012; Nef 2019). Such a varied and not always easily accessible biodiversity still poses problems for biologists in terms of censusing, thus explaining the discrepancy in the figures mentioned above. This difficulty has also arisen in the establishment of a classification system to list microalgae in distinct botanical groups (see section 1.1.2). The traditional classification initially based on the pigment composition of microalgae has established three major taxonomic groups or phyla that are commonly referred to as "green algae", "red algae" and "brown algae". Cyanobacteria, long considered as algae, were classified as "blue algae". However, the pigmentary criteria have not always been sufficient to classify the algae and report their evolutionary course. Other characteristics, such as the nature of the endosymbiosis at the origin of the chloroplast (primary or secondary endosymbiosis) (see Figure I.1), have enriched the classification by making it possible to propose new taxonomic groups. This is notably the case for the Glaucophytes group, whose chloroplast is the result of primary endosymbiosis but has retained certain important ancestral characteristics, such as the presence of a peptidoglycan wall between two membranes. This characteristic is effectively shared between these organisms and cyanobacteria (Lecointre and Le Guyader 2016). However, Glaucophytes are microalgae that are not very widespread and are found mainly in fresh water (lakes, swamps and ponds) (Lecointre and Le Guyader 2016). These organisms, which are not subject to valorization, will not be treated in the rest of this book.

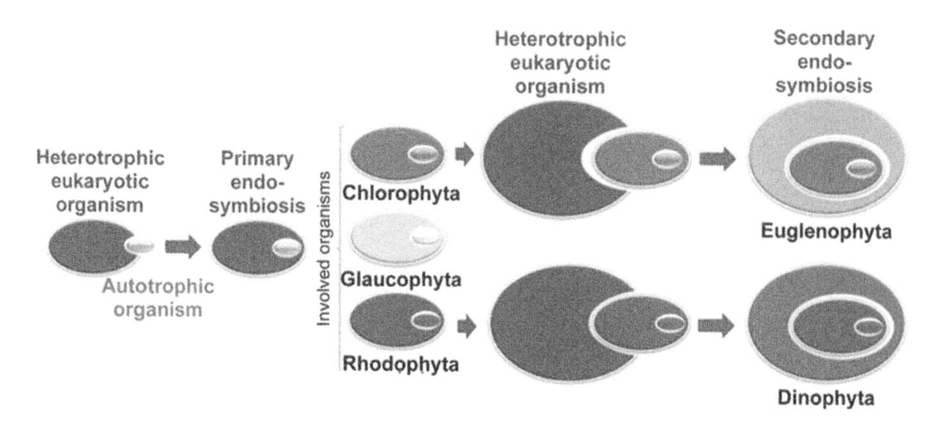

Figure I.1. *Mechanisms of primary and secondary endosymbiosis to trace the evolutionary course of certain groups of algae (source: Pouchus Y.-F.). For a color version of this figure, see www.iste.co.uk/fleurence/microalgae*

Since phylogenetic classification is mainly based on molecular criteria, such as ribosomal DNA, it has broken up the main groups previously established by the traditional classification (see Figure I.2). Nevertheless, for obvious reasons of convenience, the classifier remains attached to and still uses the denominations of red, brown or green algae to describe the algal biomass.

Microalgae show high metabolic plasticity and a natural aptitude for horizontal gene transfer. This last property is at the origin of the extremely diversified evolution of algae. These two natural characteristics of microalgae also make them valuable biotechnological aids for the production of molecules of interest, such as recombinant proteins (Cadoret *et al.* 2008) or biofuel (Maeda *et al.* 2018). Finally, microalgae, such as Chlorella, and cyanobacteria, such as Spirulina, are already used in animal and human food, even if, in the latter case, it remains more of a food supplement than a food in its own right.

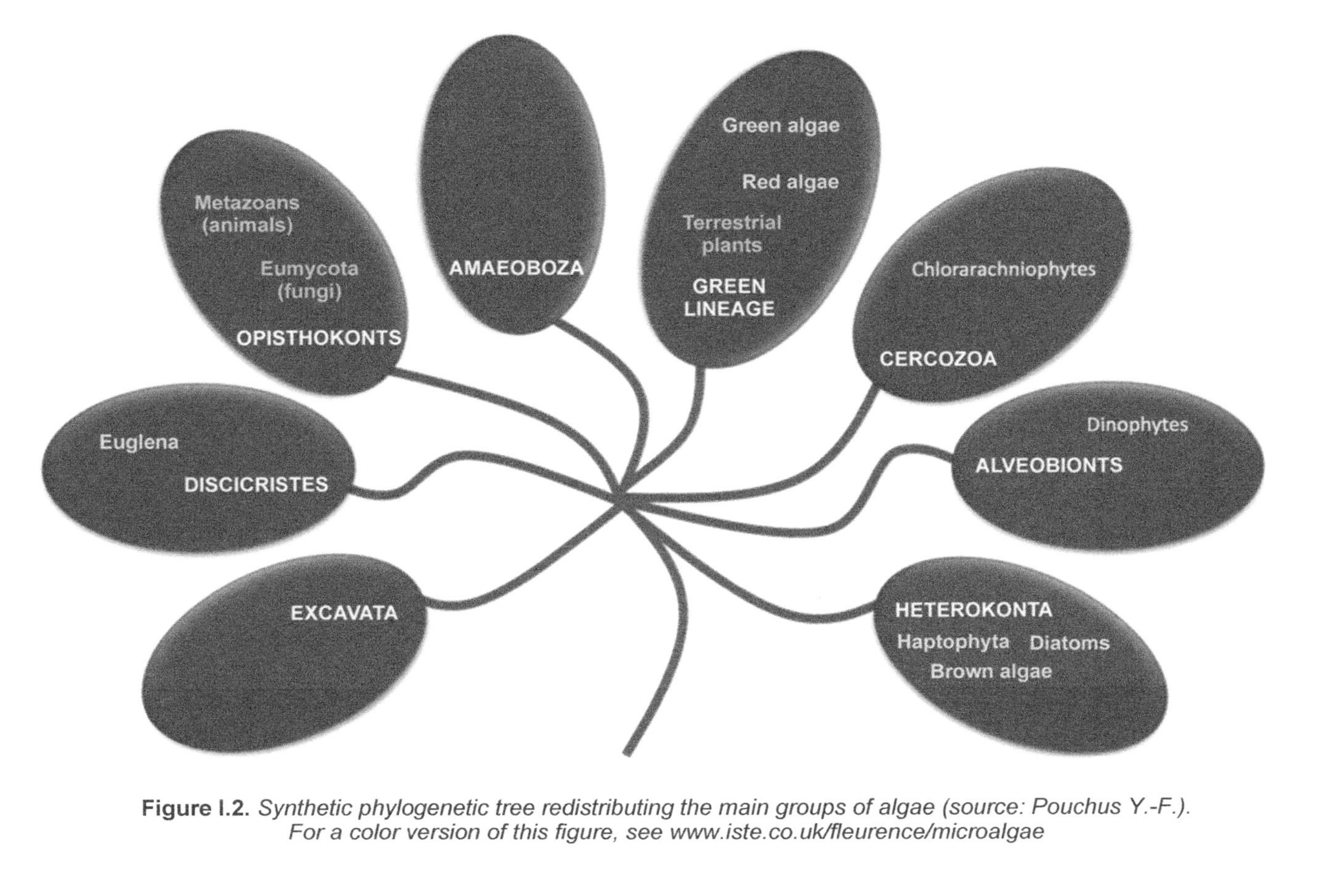

Figure I.2. *Synthetic phylogenetic tree redistributing the main groups of algae (source: Pouchus Y.-F.). For a color version of this figure, see www.iste.co.uk/fleurence/microalgae*

1

Biology and Ecology of Microalgae

1.1. Biological characteristics

1.1.1. *General characteristics*

Microalgae are mainly photo-synthetic single-celled organisms living in aquatic environments (marine, brackish, fresh water) or humid or aerial terrestrial environments (atmosphere, soils, trees, building facades, etc.) (Sharma *et al.* 2006). They can also associate together to form colonies or undifferentiated multicellular organisms. The morphology and size of microalgae vary greatly according to species and taxonomic groups.

Microalgae are eukaryotic organisms that possess the main characteristics of the vegetable eukaryotic cell. They can be flagellated as in the case of algae belonging to the genus *Chlamydomonas* (see Figure 1.1) or not (see Figure 1.2). As eukaryotes, microalgae can be distinguished from cyanobacteria, which are prokaryotic organisms with a long life span and are called "blue-green algae" or Cyanophyceae. Spirulina (*Arthrospira* sp.), a well-known representative of this group, is, therefore, a photosynthetic bacterium and not an alga, as is often stated in commercial communication (Fleurence 2018).

Microalgae are also characterized by a great morphological diversity. This is the case of diatoms, which are distinguished from other algae by the presence of a siliceous shell called a frustule and whose architecture differs according to the species considered (see Figure 1.3) (Loir 2004). This unique morphological feature is the biological signature of diatoms, which are also known as "siliceous algae". This morphological heterogeneity is also found in size since microalgae generally vary in size from less than 1 μm to 1 mm.

The smallest size, 0.8 µm, is observed for the marine species *Ostreococcus tauri* (Borowitzka 2018a).

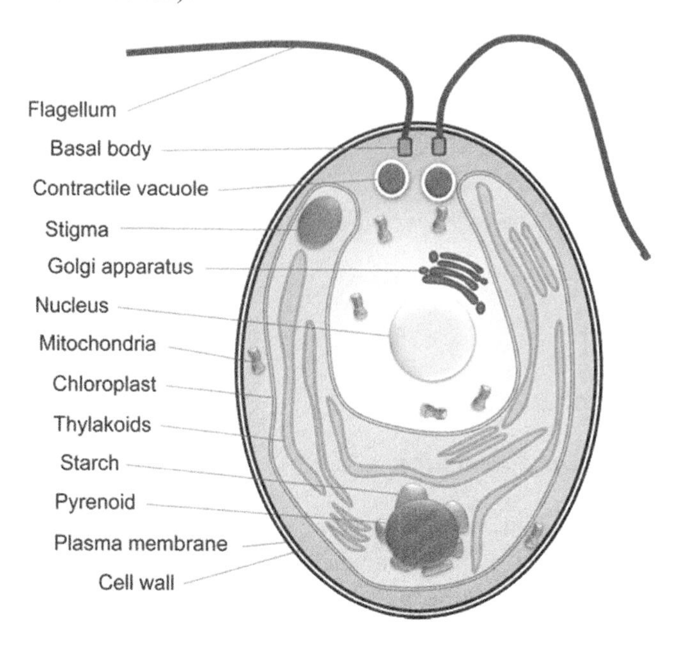

Figure 1.1. *Cellular organization of the microalga* Chlamydomonas *sp. (source: Pouchus Y.-F.). For a color version of this figure, see www.iste.co.uk/fleurence/microalgae*

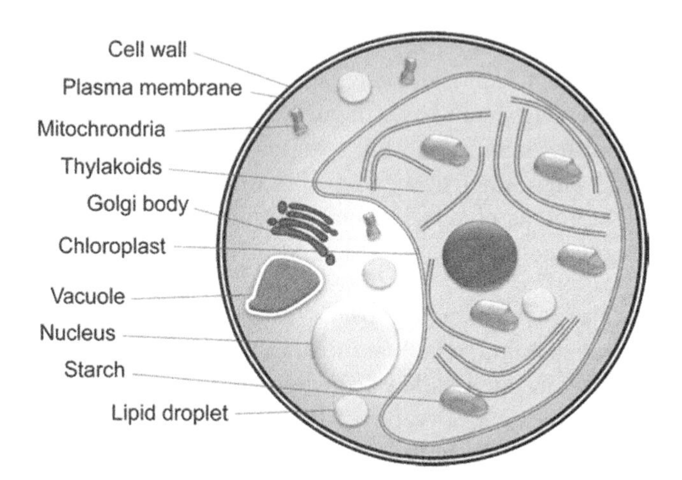

Figure 1.2. *Cellular organization of the microalga* Chlorella *sp. (source: Pouchus Y.-F.). For a color version of this figure, see www.iste.co.uk/fleurence/microalgae*

1.1.2. *The different groups in traditional and phylogenetic classification*

Like all organisms, microalgae are classified into several taxonomic groups according to traditional or phylogenetic systematics. The traditional systematics establishes botanical phyla based on the pigment position of the microalgae. The pigment criterion, according to its nature, is an agglomerating or discriminating character. Thus, all algae, micro- or macroalgae, have a common pigment: chlorophyll a. This pigment is also present in cyanobacteria and terrestrial plants. The presence of additional pigments, called supernumerary pigments, is used as a distinguishing criterion to define the main algal phyla (see Table 1.1). These phyla, of which there are three, are, respectively, the phylum of Chlorophyta (green algae), Chromophyta (golden-brown algae) and Rhodophyta (red algae). Each phylum consists of a single phylum branch, as is the case for Chlorophyta and Rhodophyta, or three separate branch, as is the case for Chromophyta (see Table 1.2). Each branch is itself divided into several classes. The number of classes may vary according to the criteria used by the different authors. This is particularly true for Rhodophyta, where complementary criteria to the pigmentary characteristics (nature of the starch, life cycle and reproduction, etc.) lead to the distinction of seven classes instead of one.

Phylum	Common pigment	Supernumerary pigments
Chlorophyta	Chlorophyll a	Chlorophyll b
Chromophyta	Chlorophyll a	– Chlorophylls c, e – Excess carotenoids (β-carotene, fucoxanthin, zeaxanthin, etc.)
Rhodophyta	Chlorophyll a	– Chlorophyll d – R or B Phycoerythrin – Phycocyanin – Allophycyanin

Table 1.1. *Pigment distribution according to the botanical phyla of microalgae (from Morançais* et al. *(2018))*

1.1.2.1. *Chlorophyta*

The phylum of Chlorophyta is characterized by the simultaneous presence of chlorophylls a and b. This group comprises 6,429 species and is divided into 11 distinct classes (Sexton and Lomas 2018) (see Table 1.2 where only the main ones are mentioned):

– the Chlorophyceae class with 3,653 species represents the main phylum taxonomic group. Microalgae belonging to the genera *Chlamydomonas*, *Volvox* and *Dunaliella* are the most representative members. The alga *Dunaliella* sp. is also valued in animal and human food (see sections 3.1 and 3.2);

– the Ulvophyceae class mainly includes macroalgae and some microscopic unicellular forms (Sexton and Lomas 2018);

– the Trebouxiophyceae class includes the genus *Chlorella*, a microalga valued in particular as a food supplement in human nutrition.

Phylum	Branch	Classes (number of species)
Chlorophyta (green algae)	Chlorophycophyta	– Chlorophyceae (3,653) – Ulvophyceae (1,725) – Trebouxiophyceae (794) – Prasinophyceae (105) – Mamiellophyceae (18)
Chromophyta (golden-brown algae)	Pheophycophyta	Pheophyceae (2,040)
	Chrysophycophyta	– Bacillariophyceae (11,000–100,000) (diatoms) – Chrysophyceae (670) – Xanthophyceae (450–600)
	Pyrrophycophyta	Dinophyceae (3,327)
Rhodophyta (red algae)	Rhodophycophyta	Rhodophyceae

Table 1.2. *Taxonomic distribution of algae according to the traditional classification enriched with the addition of secondary traits such as endosymbiotic origin plastid (based on Loir (2004) and Sexton and Lomas (2018))*

1.1.2.2. *Chromophyta*

The Chromophyta phylum is divided into three phylum branches, each of which is subdivided into several distinct classes (see Table 1.2). In the branch – of Pheophycophyta – there is one class, that of Phaeophyceae. Phaeophyceae include 2,040 species and are known as brown algae. They do not include unicellular forms, except for the presence of biflagellated spores during the reproductive cycle of these algae, the best-known members of which belong to the genera *Macrocystis*, *Laminaria* and *Fucus*.

In the Chrysophycophyta branch, there are three well-classified classes. Among them, the class of Bacillariophyceae is one of the most studied for its biological diversity and ecological importance. This class, known as diatoms, includes unicellular brown-yellow algae with sizes ranging from 2 µm to 1 mm (Loir 2004). The number of species of diatoms is estimated to be at least 11,000, but some authors estimate the number to be around 100,000 (Mann and Vanormelingen 2013). Diatoms are morphologically characterized by the presence of an external envelope of siliceous nature called a frustule. This name is derived from the Latin *frustulum*, which means "piece" or "small end" (Round *et al.* 1990). The frustule is a shell composed of two valves called the epivalve and hypovalve (see Figure 1.3). This structural element has different shapes and symmetries in different species.

The frustule is responsible for the dichotomous separation of diatoms into two distinct groups. When the latter is in the form of a disc or tube with radial symmetry, the diatoms are referred to as centric or central diatoms (see Figure 1.4). On the other hand, diatoms with a more or less elongated frustule and mainly showing bilateral symmetry are classified as pennate or pennal diatoms (see Figure 1.5).

The species *Haslea ostrearia* involved in the greening mechanism of oysters (see section 3.1) belongs to the latter group (see Figure 1.6). Diatoms are photosynthetic organisms, but some species living in light-poor environments are heterotrophic for carbon. These diatoms incapable of synthesizing chlorophyll represent less than 10 species belonging to the genera *Nitzschia* and *Hantzschia* (*Hantzschia achroma)* (Li and Volcani 1987).

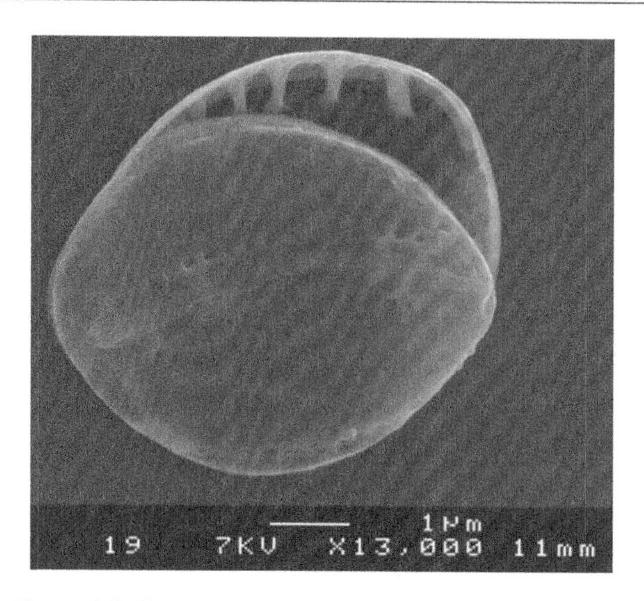

Figure 1.3. *Frustule (epivalve and hypovalve) of the diatom* Nitzschia *sp. visualized by a scanning electron microscope (SEM) (photo credit © Gaudin P., 2010)*

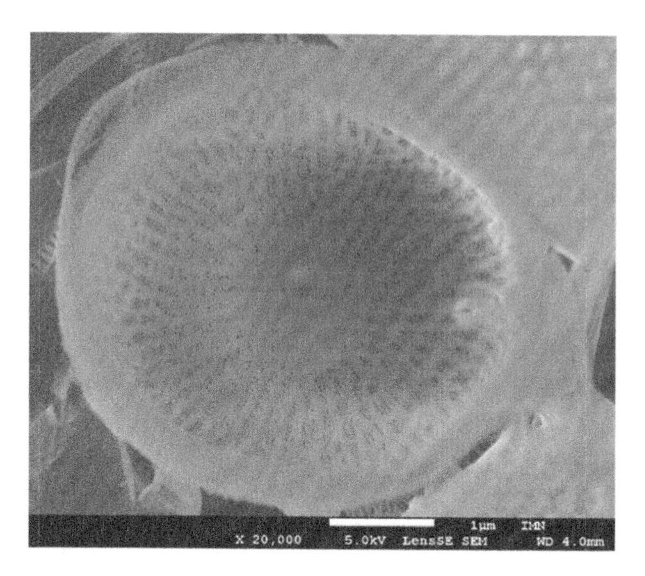

Figure 1.4. *Example of a central diatom* Skeletonema *sp. visualized by a scanning electron microscope (SEM) (photo credit © Petit A., 2010)*

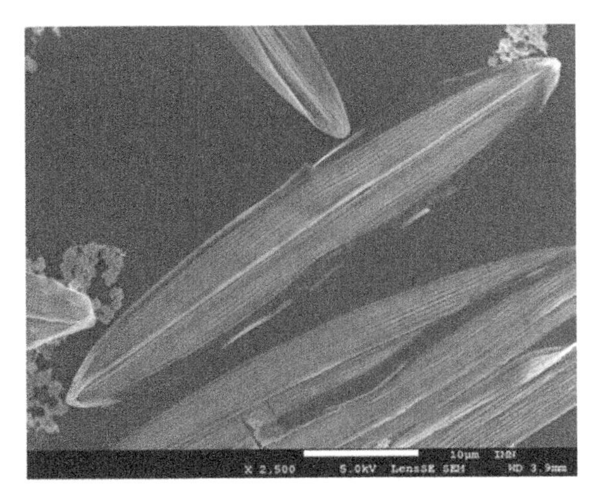

Figure 1.5. *Example of a pennate* Haslea ostrearia *diatom visualized by a scanning electron microscope (SEM) (photo credit © Petit A., 2010)*

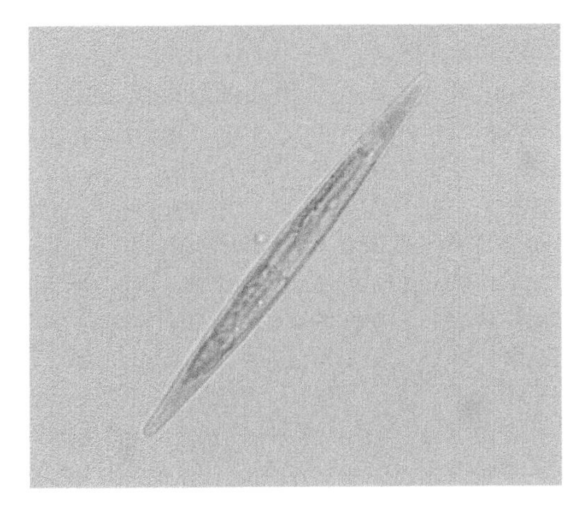

Figure 1.6. Haslea ostrearia, *pennate diatom, responsible for the greening of oysters (photo credit © Petit A., 2019). For a color version of this figure, see www.iste.co.uk/fleurence/microalgae*

Chrysophyceae or golden yellow-brown algae include about 670 species. Although autotrophic for carbon, some of them are bacterivorous and have very different morphologies, ranging from filamentous to capsular or even spherical (Sexton and Lomas 2018). The main representatives belong to the genera *Dinobryon*, *Ochromonas* and *Phaeoplaca* or *Tisochrysis*.

The Xanthophyceae, known as yellow-green algae, are well listed in the Chromophyta phylum. They do not possess chlorophyll b, which is the pigmentary criterion shared by all Chlorophyta. The yellow color is due to the presence of an excess of yellow xantho phyllum (Sharma 1986). We find in this class the algae belonging to the genera *Vaucheria* and *Chloromeson.*

The Pyrrophycophyta branch is composed of several classes, the number of which varies according to the authors and some of which are considered today as obsolete. It should simply be remembered that this branch includes the class Dinophyceae (Dodge 1985), whose organisms, known as dinoflagellates, have a very particular position in the traditional classification. These organisms have often been considered as brown algae, but their evolutionary history based on mechanisms of secondary endosymbiosis suggests that they derive from an ancestor of the Rhodophyta type (Perez 1997; Nef 2019). Moreover, dinoflagellates have the particularity, in addition to an excess of β-carotene, of possessing a red-colored pigment, peridinin (Perez 1997). This specificity is at the origin of the red coloration of waters when dinoflagellates are present in large numbers in the environment. Apart from the above-mentioned characteristics, dinoflagellates present another particularity that makes their classification delicate. Indeed, this class constitutes a mixotrophic ensemble: 50% of the species are autotrophic for carbon, the others are heterotrophic for this element (Borowitzka 2018a, 2018b). They can therefore be linked to the animal kingdom or to the plant kingdom according to this trophic characteristic. These are organisms whose size fluctuates between 10 μm and 2 mm (Spector 1984). Many species of dinoflagellates are involved in the phenomenon of "red tides". They mainly belong to the genera *Alexandrium, Dinophysis, Gymnodinium* and *Prorocentrum* (Faust and Gulledge 2002). They can produce toxins that affect invertebrates, fish and mammals. These toxins, the most well-known of which are DSP, for diarrhetic shellfish poisoning (see Figure 1.7), are at the origin of poisoning during the consumption of contaminated shellfish.

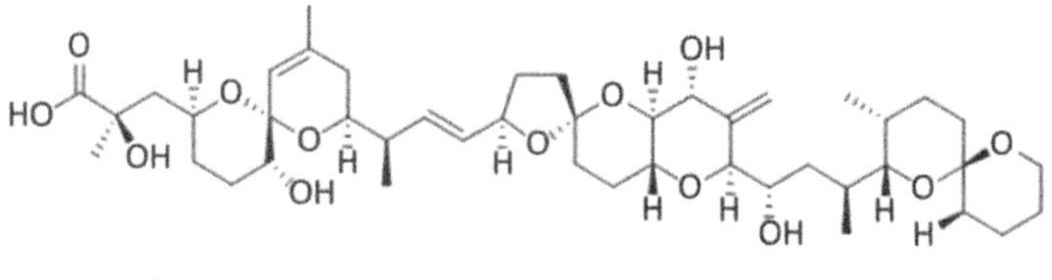

Figure 1.7. *Example of toxins produced by dinoflagellates (okadaic acid) (source: Grovel O.)*

1.1.2.3. *Rhodophyta*

On the basis of the pigment criterion, i.e. the specific presence of phycoerythrin as a supernumerary pigment, phylum has only one class (see Table 1.2). The use of additional criteria such as the nature of the starch stored in the cytoplasm, the presence of floridoside and the characteristics of the reproductive cycle led to the determination of seven classes, the main ones being Florideophyceae (6,793 species) and Bangiophyceae (196 species) (Sexton and Lomas 2018). The species *Porphyridium cruentum* (see Figure 1.8), an alga belonging to the latter class, is the subject of numerous evaluations, in particular as a cell factory for the production of exopolysaccharides, pigments or enzymes (super oxide dismutase) (see section 4.2).

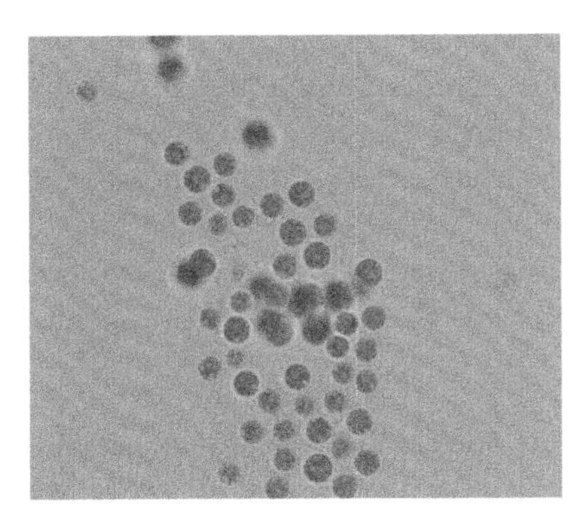

Figure 1.8. Porphyridium cruentum *(photo credit © Petit A., 2019). For a color version of this figure, see www.iste.co.uk/fleurence/microalgae*

The traditional classification, although complex, classifies microalgae into three distinct groups: green algae, brown algae and red algae. This dichotomous division proves practical and useful for the valorization of this resource, whether for producers, processors or consumers.

1.1.2.4. *The green lineage and other clades*

Microalgae, like all organisms, can also be classified according to phylogenetic systematics. This approach, whose scientific interest is undeniable, is not easy to use in the framework of a valorization approach. Indeed, it is not based on the composition of pigments, molecules that can be valorized, but on genomic or morphological criteria. It is based, in particular, on the construction of the evolutionary history of algae from molecular criteria, such as DNA sequences coding for ribosomal subunits or those of ribosomic RNA. It is also based on the mechanisms of endosymbiosis at the origin of the algal cell and certain morphological criteria, such as the presence of a single flagellum (unikont) or a double flagellum (bicont) in unicellular forms of algae (microalgae or spores and gametes of macroalgae) (Lecointre and Le Guyader 2016). This approach leads to the isolation of certain groups of algae from others by associating them, for example, with unicellular organisms with animal affinity such as protozoa (Lecointre and Le Guyader 2016). In this new context, the dichotomous plant/animal division appears, according to some authors, outdated (Selosse 2006).

For phylogeneticists, algae do not constitute a homogeneous group from an evolutionary point of view. The use of molecular markers such as ribosomal DNA or ribosomal RNA has shown that these organisms do not constitute a monophyletic group. Within the eukaryotic phylogenetic tree, red and green algae are grouped in the "green line" group, which includes terrestrial plants (see Figure I.2). Dinoflagellates, microalgae initially grouped in the phylum of Chromophyta and the branch of Pyrrophycophyta (Spector 1984), are now listed within the Alveolobiontes or Alveolates group, which includes protozoa, such as ciliates, or sporozoan, or apicomplexa (Bhattacharya *et al.* 1992).

Diatoms, Haptophyta and Phaeophyceae, are associated with oomycetes (Aritzia *et al.* 1991) within the group of Heterokonta (Selosse 2006). Haptophyta, whose plastids have an endosym biotic rhodophyta origin and whose main characteristic is the presence of an organelle acting as a flagellum (haptonema) (Sexton and Lomas 2018), were thus isolated from their initial phylum. This new classification shows that the pigment character initially chosen to classify the algae was not a sufficiently relevant criterion to retrace the evolutionary history of the algae.

Nevertheless, this criterion is still relevant to distinguish the major botanical groups of algae that are being valorized, whether in the field of biotechnology or in that of human or animal nutrition.

1.1.3. *The special case of cyanobacteria (Cyanophyceae)*

Cyanobacteria have long been considered microalgae and classified as Cyanophyceae or "blue algae" or "blue-green algae". These organisms belong to the prokaryotic lineage, which is not the case for algae, which are all eukaryotic organisms. Cyanobacteria are the most important group of photosynthetic bacteria. The size of these organisms can vary between 1 µm and 10 µm. Cells can be isolated or form cell colonies of various shapes. In particular, they can produce filamentous excretions or trichomes. The cells are sometimes in close contact in a trichome, as is the case for the genus *Oscillatoria* (Lecointre and Le Guyader 2016). Colonies can also take on a spiral shape (see Figure 1.9).

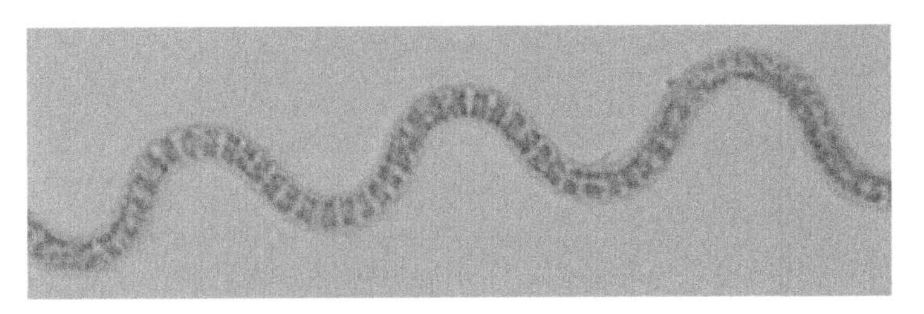

Figure 1.9. Spirulina platensis (Arthrospira platensis) *(photo credit © Jubeau S., 2019). For a color version of this figure, see www.iste.co.uk/fleurence/microalgae*

This is particularly the case of the species *Spirulina platensis*, which takes its name from this particular morphology. This species, renamed *Arthrospira platensis*, is currently valued as a food supplement in human nutrition (see section 3.2). Cyanobacteria constitute an important taxonomic group from the point of view of the number of species, although this number may vary according to the authors (Guiry 2012; Sexton and Lomas 2018). It is reported to be between 3,000 and 8,000 species, with the number of 8,000 being an estimate and not an exhaustive census (Guiry 2012) (see Table 1.3). In the traditional classification, cyanobacteria were listed in the phylum of Cyanophyta, the branch of Cyanophycophyta with a single class of Cyanophyceae. The phylogenetic classification revisited this group by integrating it into the field of eubacteria and dividing it into five different taxa (see Table 1.4) (Lecointre and Le Guyader 2016).

Cyanobacteria (Cyanophyceae or blue algae)	Number of species listed	Estimated number of species	Number total of species
According to Guiry (2012)	3,000	5,000	8,000
According to Sexton and Lomas (2018)	4,258	–	–

Table 1.3. *Assessment of the number of cyanobacteria species according to some authors*

Cyanobacteria	Some examples of species
Chroococcales	*Chroococcus membraninus* *Cyanobacterium stanieri* *Cyanobium gracile* *Cyanocystis violacea* *Dactylococopsis linearis* *Prochlorococcus marinus* *Synechocystis salina* *Gloeobacter violaceus*
Nostocales	*Anabaena variabilis* *Raphidiopsis mediterranea* *Nostoc commune*
Oscillatoriales	*Arthrospira platensis (Spirulina platensis)* *Oscillatoria princeps* *Spirulina labyrinthiformis* *Symploca atlantica* *Planktothrix agardhii*
Pleurocapsales	*Pleurocapsa minor* *Dermocarpella incrassata* *Chroococcopsis gigantea*
Stigonematales	*Stigonema tomentosum* *Nostochopsis lobatus* *Fischerella thermalis*

Table 1.4. *Taxonomic distribution of cyanobacteria according to phylogenetic classification (Lecointre and Le Guyader 2016)*

The cyanobacterium *Nostoc*, like *Arthrospira*, is used in human food (Chisti 2018).

1.2. Ecological features

Microalgae are present in many environments, whether aquatic, terrestrial or atmospheric. In the aquatic environment, marine microalgae can be distinguished from freshwater microalgae or freshwater algae. Marine and freshwater algae can live free in the water column or reside on the bottom, developing a thin layer called a biofilm. Seaweed in the water column is called phytoplankton, and seaweed on the bottom is called microphytobenthos. Phytoplankton can be subdivided into several categories according to the size of the organisms (see Table 1.5).

Category of phytoplankton	Size
Picoplankton	0.2 μm < x < 2 μm
Nannoplankton	2 μm < x < 20 μm
Microplankton	20 μm < x < 200 μm
Macroplankton	200 μm < x < 2,000 μm

Table 1.5. *Distribution of different categories of phytoplankton by size (from Iltis (1980))*

In fresh water, plankton is called potamoplankton. This name comes from the ancient Greek *potamos* or river, which etymologically means "river plankton". Whether marine or fresh water, microalgae can live in association with other organisms, such as cyanobacteria and fungi, to form a mixture called periphyton. Periphyton can colonize the surface of other organisms or inert media, such as rocks or submerged surfaces like plastics or metals.

1.2.1. *Marine microalgae*

Marine phytoplankton is responsible for half of the oxygen, the rest being generated by terrestrial plants and more particularly by the Amazonian forest (see Figure 1.10). It also acts as a climate regulator because of its capacity to sequester atmospheric CO_2. Phytoplankton also contributes to other biogeochemical cycles in the ocean. In particular, it participates in the nitrogen cycle. Cyanobacteria fix molecular nitrogen (N_2), and microalgae absorb other forms of available nitrogen, such as nitrates (NO_3^-), nitrites (NO_2^-) or ammonium (NH_4^+) (Nef 2019). It is also involved in the phosphorus and sulfur cycles. Dinoflagellates are among the main producers of dimethylsulfo niopropionate or DMSP, which can be transformed into dimethylsulfide (DMS), via the biological activity of certain organisms, before being converted into sulfuric acid in the atmosphere (Lovelock *et al.* 1972; Nef 2019). The production of sulfuric acid, an agent of cloud condensation, and of DMS that would modify the reflective capacity of clouds or albedo (Charlson *et al.* 1987) shows that the role of phytoplankton in climate regulation is not limited to its action as a carbon dioxide trap. Finally, the diatoms that represent the main group of microalgae in phytoplankton participate in the silicon cycle in the oceans through the synthesis of their frustule, which requires the contribution of silica. The organisms making up phytoplankton live in the photic zone, which is defined as the area of the ocean into which sunlight penetrates up to a limit value of 1%. Below this limit, one leaves the photic zone and enters the aphotic zone. However, the depth of the photic zone varies according to the turbidity of the water. In particular, it can go down to 200 m in the absence of turbidity. Phytoplankton is also the first link in the food chain. As such, it is considered to be the primary producer of the entire oceanic food web. Phytoplankton is mainly composed of cyanobacteria such as *Prochlorococcus marinus*, diatoms such as *Phaeodactylum tricornutum* and dinophyceae (Sardet 2013).

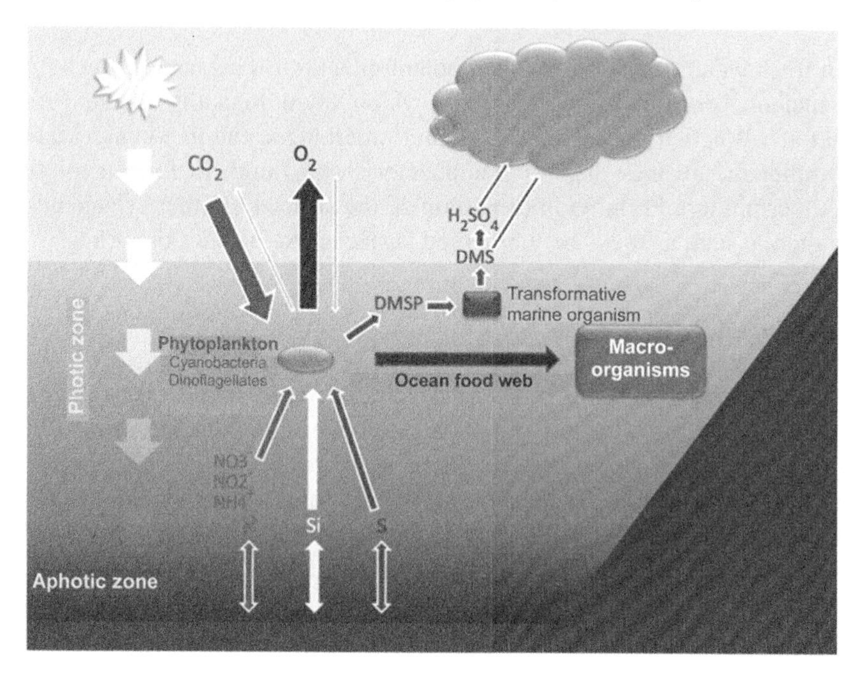

Figure 1.10. *Role of phytoplankton in the biogeochemical cycle of the ocean (source: Pouchus Y.-F.). For a color version of this figure, see www.iste.co.uk/fleurence/microalgae*

Diatoms, the main group of microalgae in the marine environment, can be planktonic (phytoplankton) or benthic (microphytobenthos) (see Table 1.6). Some benthic species can be re-suspended by currents and support life in the open water. These species are called tychoplankton.

Species	Planktonic diatom	Benthic diatom	Tychoplanktonic diatom
Chaetoceros sp.	+		
Odontella aurita	+		
Pseudo-nitzschia pungens	+		
Nitzschia angularis		+	
Nitzschia reversa		+	
Nitzschia bilobata		+	
Amphora inflexa		+	+

Table 1.6. *Some examples of planktonic, benthic and tychoplanktonic marine diatom species (from Loir (2004))*

1.2.2. *Microalgae in brackish and freshwater environments*

Microalgae are also present in brackish and fresh waters. Brackish waters are intermediate environments with respect to the marine and freshwater environment since the salinity varies between 1 and 10 g/L of salt (Loir 2014). In both types of environments, centric or pennate diatoms are found (see Table 1.7).

Species	Centric diatom	Pennate diatom
Melosira moniliformis	+	
Melosira nummuloides	+	
Fragilaria pulchella		+
Cocconeis scutellum		+
Anomoeonis sphaerophora		+
Diploneis didyma		+

Table 1.7. *Examples of common centric or pennate diatom species in brackish water (from Loir (2014))*

Some species can be found in both marine and brackish waters. This is particularly the case of the *Cocconeis scutellum* species, which is an organism that is very widespread on our coasts and which shows a very great adaptation to variations in salinity.

Freshwater diatoms are significantly less numerous than those living in the marine environment. Centered diatoms are represented in our rivers or lakes by only about 30 species (Loir 2014). Freshwater diatoms are generally benthic and can be found attached to mud, pebbles, submerged structures or even organic supports such as macrophyta, other diatoms or driftwood. The species *Melosira varians*, a diatom centered and very common in watercourses, lives attached to stones, algae or even submerged wood. This species can also be planktonic.

Apart from diatoms, fresh water contains many species of microalgae belonging to the group of Chlorophyta or related. These are mainly species belonging to the genera *Chlamydomonas*, *Pandorina*, *Volvox* and *Chlorella* (Greenhalgh and Ovenden 2009). Finally, there are also euglena belonging to the phylum of Euglenophyta, which derive by secondary endosymbiosis from the phylum of Chlorophyta (see Figure I.2).

Fresh water also contains cyanobacteria. Some species belonging to the genus *Arthrospira*, formerly *Spirulina*, have been consumed by people living near the lakes of the Mexico Valley (Sanchez *et al.* 2003). Cyanobacteria are also the cause of invasive phenomena or blooms in rivers, lakes or reservoirs, following the artificial enrichment of the environment with nutritive salts (Anderson *et al.* 2002). Other phenomena such as climate change predict the development of these cyanobacterial bloom phenomena in the future (Carey *et al.* 2012).

1.2.3. *Microalgae in terrestrial and aerial environments*

The presence of microalgae outside the aquatic environment is a well-known reality. Microalgae and cyanobacteria grow on wet fronts and fix the surrounding CO_2 while producing oxygen (see Figure 1.11(a) and (b)). In particular, they have replaced lichens on walls and tree trunks in highly polluted cities (Gudin 2013).

Recovering the conditions of the primitive earth in urban environments loaded with CO_2 and hydrogen sulfide, cyanobacteria and microalgae thus appear as potential depollution auxiliaries (Gudin 2013). Certain microalgae, such as *Porphyridium cruentum* nesting in damp churches, could, via the excretion of their purple pigment, phycoerythrin, be at the origin of the phenomenon of bleeding statues (Fleurence 2018). The atmosphere is not a normal habitat for microalgae. The algae present in the air are of various shapes and sizes (from 1 to 150 μm). The atmospheric algal flora is mainly composed of Chlorophyta, Chromophyta (diatoms, xanthophyceae) and cyanobacteria (Sharma and Rai 2011) (see Table 1.8).

a)

b)

Figure 1.11. *Examples of facades colonized by microalgae (photo credit © Fleurence J., 2019). For a color version of this figure, see www.iste.co.uk/fleurence/microalgae*

This biological diversity varies according to climatic conditions, geographical and topographical characteristics. For example, in tropical regions, the biodiversity of the aerial algal flora is richer than in other climatic regions. Cyanobacteria, in particular, dominate the tropical atmospheric flora. In contrast, Chlorophyta are the main organisms of this flora in temperate regions (Sharma and Rai 2011).

Species	Cyanobacteria	Chlorophyta	Xanthophyceae (Chromophyta)	Diatoms (Chromophyta)
Anabaena sp.	+			
Arthrospira sp.	+			
Nostoc sp.	+			
Spirulina sp.	+			
Chlamydomonas sp.		+		
Chlorella sp.		+		
Nannochloris sp.		+		
Microspora sp.		+		
Heterococcus sp.			+	
Vaucheria sp.			+	
Nitzschia sp.				+
Melosira sp.				+
Navicula sp.				+

Table 1.8. *Examples of genera constituting the algal aerial flora (from Sharma and Rai (2011))*

Microalgae and cyanobacteria are transferred to the layers of the atmosphere by a process called aerosolization, which transforms them into fine algal particles (see Figure 1.12). They are released from terrestrial or aquatic sources, as well as from vegetation or building facades and roofs by the action of wind and updrafts (see Figure 1.12).

Microalgae are varied unicellular organisms living in different environments or habitats. They show a great biological, biochemical, genetic and ecological adaptability. These properties are notably at the origin of their development both in the field of biotechnology and in that of nutrition.

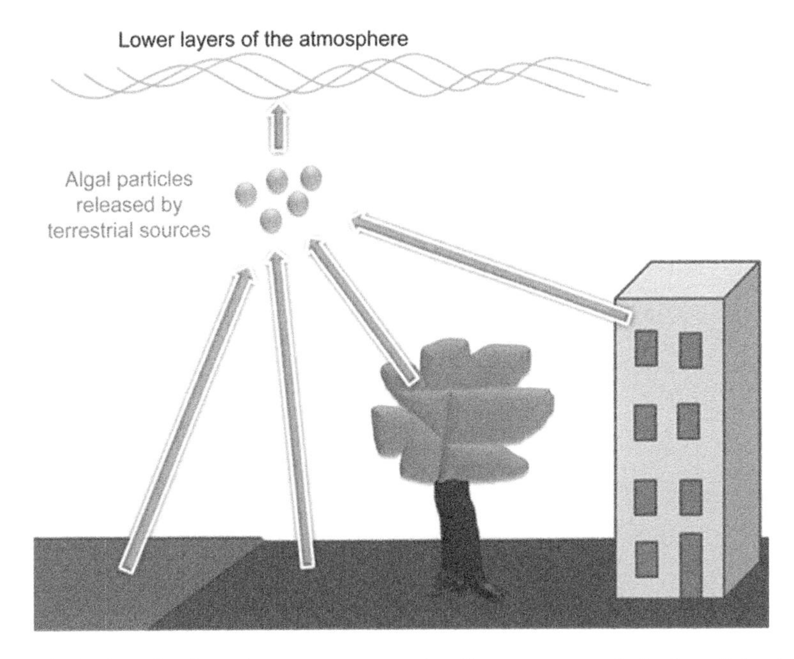

Figure 1.12. *Aerolization phenomenon causing the presence of algal particles in the layers of the atmosphere (source: Pouchus Y.-F., according to Sharma and Rai (2011)). For a color version of this figure, see www.iste.co.uk/fleurence/microalgae*

2

Production Techniques

The production of microalgae can be facilitated by harvesting in a natural environment or by culture in an artificial environment. Two modes of culture are possible: open-system culture and closed-system culture. Open-system production is a cultivation method that dates back to the 1950s: the algae are produced in open ponds or tanks. Closed-system culture is carried out in closed reactors in the form of bags, conical cylinders, tubular networks or alveolar plates.

2.1. Production by harvesting in the natural environment

Cyanobacteria of the genera *Arthrospira* (*Spirulina*), *Nostoc* and *Aphanizomenon* have been used as food for thousands of years (Spolaore *et al.* 2006). Historically, *Spirulina* was consumed by the lakeside populations of the Kanem region in Chad and by the Aztec people living on the shores of Lake Texcoco in Mexico (Jarisoa 2005). *Spirulina* shows a relatively large size between 200 and 500 μm. When sufficiently concentrated in water, it forms a colorful bloom that is easily harvested using very fine mesh nets or a container (see Figure 2.1). After draining the harvest and drying, a paste is then obtained, which serves as a basis for the preparation of galettes[1]. These galettes are still sold in some local markets in Chad under the name "*dihé*" (see section 3.2.1).

Cyanobacteria of the genus *Nostoc* can form copious macro-colonies, sometimes in the form of unbranched filaments or trichomes (Jarisoa 2005).

1 A term used in French cuisine to refer to various types of crusty cakes.

These colonies will be observed in ponds or on wet soils. In ponds, they are colonies of *Nostoc pruniforme* called "pond eggs". On calcareous soils, dehydrated colonies swell in volume by turgidity after the passage of rain and form large filamentous mats that cover and protect poor soils. These organic structures called "moon spittle" or "devil's spittle" are easily harvestable and consumed by the local population. However, their intensive harvesting is at the origin of the erosion of many limestone soils in China, and the authorities prohibit their removal in many regions.

Another cyanobacterium belonging to the genus *Aphanizomenon* is collected from the surface of Klamath Lake in the state of Oregon in the United States. This lake is the only source used to supply this organism which is valued as a dietary supplement (see section 3.2.2).

The microalga *Dunaliella salina*, present in the salt marshes of the Camargue, proliferates until it colors the entire surface of the water red. It will then form a red mud easily harvestable by the salt workers. The biomass thus harvested is mainly used in a range of cosmetic products (see Figure 2.2).

Figure 2.1. *Harvesting of the* Spirulina *bloom on the shores of Lake Chad (photo credit © FAO/Marzot M., 2010). For a color version of this figure, see www.iste.co.uk/fleurence/microalgae*

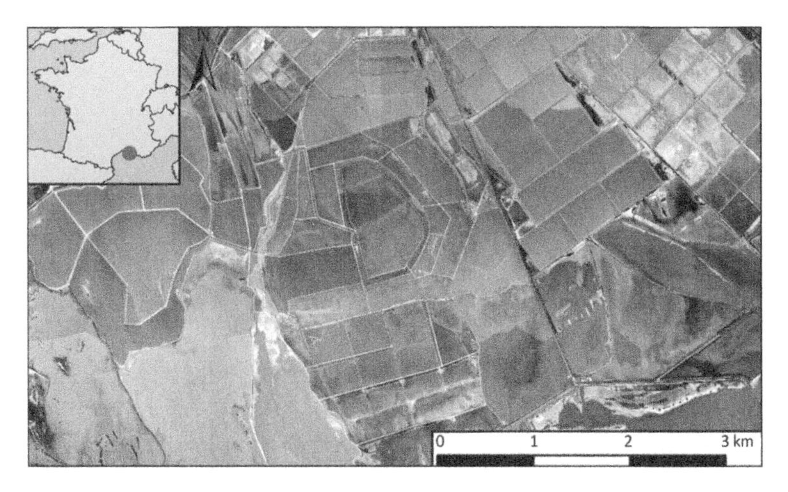

Figure 2.2. *Proliferation of the alga* Dunaliella salina *in salt marshes in the Camargue (photo credit © IGN, modified Brunier G., 2020). For a color version of this figure, see www.iste.co.uk/fleurence/microalgae*

Today, cyanobacteria and microalgae are cultivated and are no longer harvested in the natural environment, except for a few local or rather special applications.

Microalgae and cultured cyanobacteria are produced by cultivation in open or closed systems. In both systems, algal biomass can be produced in a continuous or discontinuous mode. Continuous mode culture involves the continuous supply of nutrients during cultivation. In contrast, in discontinuous mode, nutrients are supplied at the start of the culture, and no intervention on the medium or biomass is made until the end of the culture. This mode of culture is also known as *batch* culture.

Open-system culture is generally associated with low or medium added values (e.g. fodder algae, feed supplements). Closed-system culture in batch mode is recommended for obtaining axenic strains. Continuous closed-system culture is intended for applications with high economic value, such as the production of molecules with high added value (e.g. pigments, enzymes, polysaccharides). This continuous culture is based on a tubular network or photobioreactor technology.

2.2. Production by culture in open systems

2.2.1. *Production in open basins*

The production of microalgae in an open system takes place in large artificial basins or ponds. In Australia, this production mainly concerns the alga *Dunaliella salina*, which is cultivated for its high concentration of β-carotene (Borowitzka 2005).

These basins consist of a large body of shallow water (20–30 cm) and can cover an area from 1 to 200 ha. This type of culture is well adapted to *D. salina*, as this species grows in very high salinity environments, and an increase in saline concentration, for example, by natural evaporation of water, is sufficient to avoid contamination of the culture by other microorganisms. This extensive culture is usually not subjected to water agitation. The lack of water circulation, however, leads to many disadvantages, such as sedimentation of the microalgae, a concentration gradient of the culture from surface to bottom, or a temperature gradient. This heterogeneous distribution of the cultivated biomass over time can end up limiting the performance of the productive system. Indeed, for *D. salina*, the maximum biomass concentration obtained is 0.1 g/L without water agitation, compared to 1 g/L with water agitation (Borowitzka 2005). According to Borowitzka, a current velocity of 20–30 cm/s is necessary to avoid problems of sedimentation of the culture.

The increase in the speed of agitation, even if it seems to have a positive effect on growth, however, induces an additional economic cost for production. This low-cost cultivation method is not feasible in all regions, as it requires large, accessible areas and an exceptional climate of sunshine and low rainfall.

Circular basins with an axial arm for water mixing are also used for the production of microalgae in open systems. These basins are identical in design to those used for the treatment of waste water in sewage treatment plants. This type of rotating arm basin is used for the production of *Chlorella* in Japan and Taiwan (Borowitzka 2005).

2.2.2. *Production in open raceway-type basins*

The open-system culture is also based on the use of pools called raceways. These ponds have a geometric shape resembling a circuit or racetrack, and the water is circulated by a paddle wheel (see Figure 2.3).

Figure 2.3. *Raceway-type basin (photo credit © Braud J.-P., 2020). For a color version of this figure, see www.iste.co.uk/fleurence/microalgae*

This type of basin has been used since the 1950s for the production of algal biomass or wastewater treatment (de Godos *et al.* 2014). Nearly 95% of the production of commercially available microalgae is carried out using raceway-type ponds (Duarte-Santos *et al.* 2016).

Species	**Type of application**
Odontella aurita	Human food
Skeletonema costatum	Cosmetics
Phaeodactylum tricornutum	Cosmetics

Table 2.1. *Species cultivated in raceways for food and cosmetic applications by the company Inovalg (from Braud (2020))*

This cultivation method is particularly well developed for the production of *Arthrospira* sp. (*Spirulina*), *Dunaliella* sp. or *Haematococcus* sp.

(Borowitzka 2005). In France, the company Inovalg produces three species for human food or cosmetics using this cultivation method (see Table 2.1).

In a raceway system, nutrients and water are supplied continuously, and the biomass is removed as it is needed to avoid shading of the upper algal layer on the lower layers. This type of technology is subject to many constraints that can limit its effectiveness. Among these is the limitation of the transfer of CO_2 in gaseous form to the liquid culture medium of the microalgae. Under phototrophic conditions, only 5% of the carbon necessary for biomass development is directly transferred from the atmosphere to the culture medium (de Godos *et al.* 2014). Carbon availability thus appears to be a major limiting factor for this mode of culture. This disadvantage is generally mitigated by the injection of CO_2 into the culture medium. Carbon dioxide is injected into the sump that supplies water to the ponds and not into the pond spans (Stepan *et al.* 2002). Depending on the degree of purity of the CO_2 injected, the depth of the sumps influences the efficiency of the transfer between the gaseous and liquid media. The introduction of pure CO_2 into a liquid medium circulating at a velocity of 20 cm/s from a sump 1.5 m deep generates a carbonaceous transfer efficiency of 95% between the two media (Stepan *et al.* 2002). The use of a gas low in CO_2 (15%) in a liquid circulating at a velocity of 30 cm/s requires the use of a 13.4 m sump to ensure a transfer efficiency of 90% of the CO_2 between the two media (Stepan *et al.* 2002).

The CO_2 content of the injected gas and the speed of the circulating liquid are not the only parameters that can influence the carbon transfer between the media and the biomass. The gas flow rate is an important criterion and can counterbalance the CO_2 content of the injected gas. Thus, a gas with a 14% CO_2 content and injected at a flow rate of 150 L/min sees its transfer efficiency to the liquid medium reduced to 67%, and only 32% of the carbon will be fixed by the biomass (Duarte-Santos *et al.* 2016). Conversely, on the same type of culture, gas injection at CO_2 contents ranging from 2 to 6% and gas flow rates between 75 and 100 L/min allows a carbonaceous gas/liquid transfer efficiency of nearly 95%. In such a system, 85% of the carbon injected into the medium is assimilated by the biomass (Duarte-Santos *et al.* 2016). The "design" of the installations, the degree of purity of the CO_2 injected and especially the gas flow appear to be the determining criteria to ensure optimal carbon transfer between gaseous/liquid media and biomass. CO_2 injection to raceway cultures is also necessary to maintain pH during the biomass production phases (see Figure 2.4).

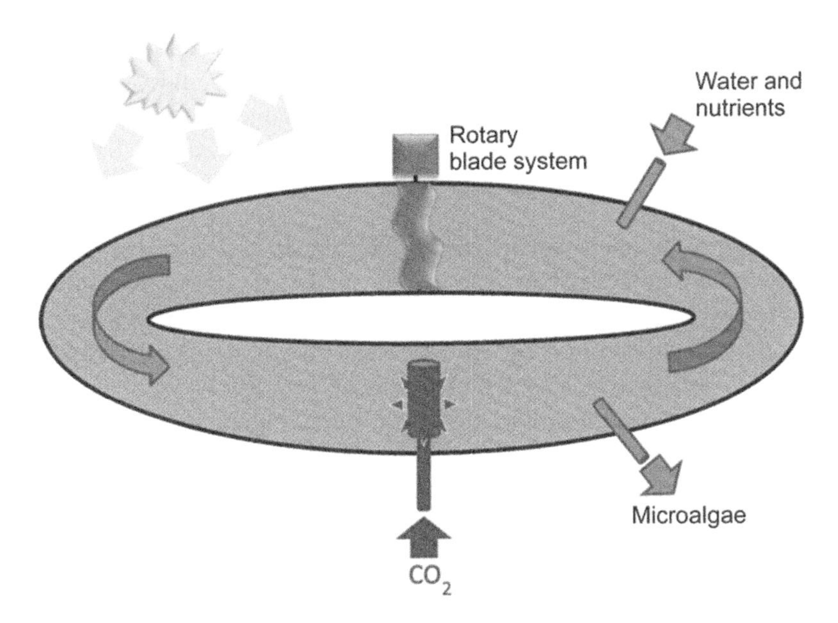

Figure 2.4. *Representation of the open culture system of microalgae in a raceway-type basin (source: Pouchus Y.-F.). For a color version of this figure, see www.iste.co.uk/fleurence/microalgae*

Raceway installations consist of ponds that can reach a surface area of 1 ha with a depth of 20–30 cm. These basins have the advantage of being simple and have a low construction cost. However, they can only be used for the culture of strains of microalgae or cyanobacteria that are not very or moderately sensitive to contamination by other microorganisms. The theoretical production of such ponds is estimated at 40 g of biomass per day per m^2, leading to a theoretical annual production of 146 tonnes of biomass per year per ha (Brennan and Owende 2010). For *Spirulina*, some authors estimate the dry annual biomass produced at 60 or 70 tonnes per hectare for installations located in areas with a subtropical climate with low cloud cover (Richmond *et al.* 1990). However, the productivity figures recorded are much lower than those theoretically calculated and differ greatly according to the species cultivated. For *Spirulina*, the maximum amount of biomass obtained in a raceway culture system is 21 $g/m^2/d$ (Vonshak and Guy 1992). This same productivity falls to 13 $g/m^2/d$ for *Chlorella* (Hase *et al.* 2000).

Raceway culture has also been the subject of a semi-discontinuous procedure similar to batch culture (Richmond *et al.* 1990). This type of culture has been optimized for the *Spirulina platensis* species for long-term cultivation. In this case, the main parameters to be improved were the composition of the nutrient medium supplied, the concentration of *Spirulina* in the culture (0.4 g/L) and especially the culture renewal rate between 40 and 60% (Richmond *et al.* 1990).

The culture in open-system and, more particularly, in *raceway*-type ponds remains, however, an interesting mode of production from an economic point of view. It has the major disadvantage of exposing the cultures to contamination by an external microorganism. While this possibility is relatively rare for the culture of *D. salina* species, given the hypersaline living conditions of this alga, other organisms of interest are more likely to be contaminated, such as *Chlorella* sp. or *Arthrospira* sp. (*Spirulina*) are more sensitive to such a contamination phenomenon. For *Spirulina*, contamination of raceway cultures by *Spirulina minor* or *Chlorella* sp. would generate a yield loss in biomass production of 15–20%.

Another disadvantage of raceway production is the difficulty in obtaining standardized biomass. However, this disadvantage can be avoided by regular sampling of the biomass produced. Raceway technology is therefore well adapted to the production of biomass for animal or human food in view of its moderate economic cost. However, its use is less relevant for biotechnological purposes, such as the production of molecules of interest such as therapeutic active ingredients, enzymes or biofuel.

There is a variation to *raceway* systems that differs from the latter in the shape of the pools and the type of application that results from this mode of cultivation. It is a technology based on the use of rectangular basins and whose water is stirred by a paddle system. In this system, microalgae (*Chlorella* sp., *Scenedesmus* sp.) or cyanobacteria (*Arthrospira* sp.) previously seeded will allow the production of an algal biomass and the treatment of wastewater. The algae provide the oxygen necessary for the purifying bacteria to function and capture the mineral elements present in the environment. The algal biomass produced in such a system can be used for low value-added applications, such as animal nutrition, but cannot be used for human food (Borowitzka 2005; Becera-Celis 2009).

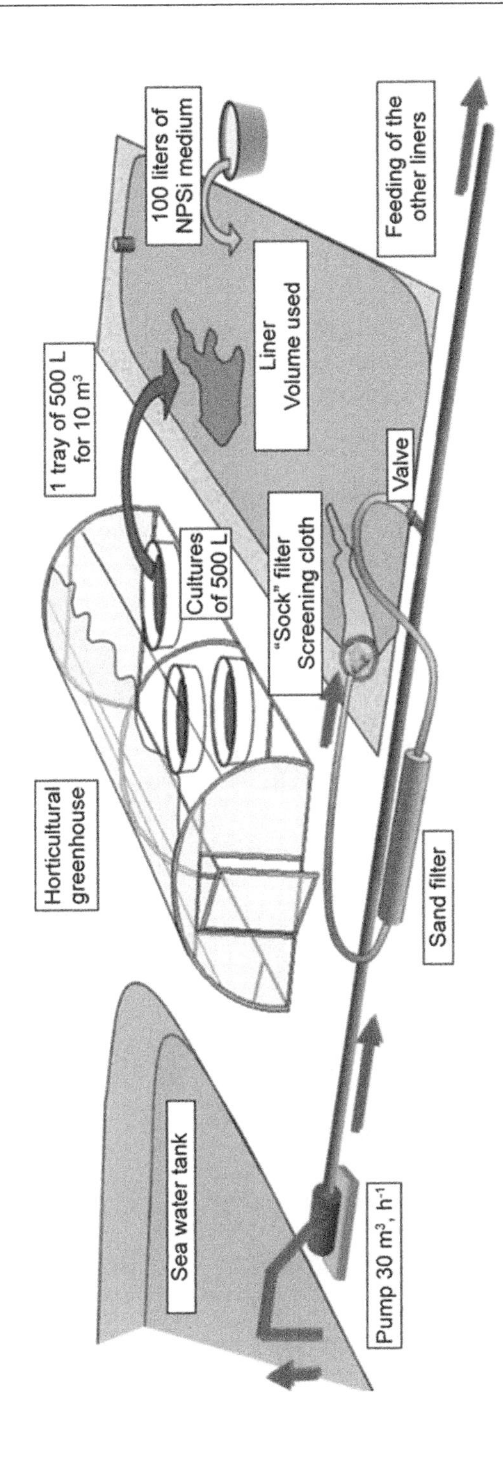

Figure 2.5. Haslea ostrearia *microalgae culture and greening facility of the oyster* Crassostrea gigas *(source: (Turpin 1999)). For a color version of this figure, see www.iste.co.uk/fleurence/microalgae*

2.2.3. *Open-tank production*

The production of algal biomass in tanks of limited volume (500 L) is also an open-system culture method. In this method, algal inoculum and nutrients are added at the beginning of the culture, and the production method is called *batch*.

This type of culture has been used for the production of the microalga *Haslea ostrearia* with a view to supplying it as fodder algae in the feed of the oyster *Crassostrea gigas* (Turpin 1999). In this production system, the algal biomass production unit is located close to the oyster refining basin or greening basin (see Figure 2.5).

The algal biomass cultured in the tanks is then transferred to ponds to be consumed by the oysters and participate in the greening phenomenon of the bivalve mollusk (see Figures 2.5–2.8).

Figure 2.6. *Growing tray of the microalga* Haslea ostrearia *(500 L) under horticultural glasshouse (photo credit © Turpin V., 1999) . For a color version of this figure, see www.iste.co.uk/fleurence/microalgae*

Figure 2.7. *Oyster greening basin with the addition of* Haslea ostrearia *culture (photo credit © Turpin V., 1999). For a color version of this figure, see www.iste.co.uk/fleurence/microalgae*

Figure 2.8. *Oyster verdie (*Crassostrea gigas*) after consumption in the greening basin of the microalga* Haslea ostrearia *(photo credit © Turpin V., 1999). For a color version of this figure, see www.iste.co.uk/fleurence/microalgae*

2.3. Production by culture in a closed system

The production by culture in a closed system can be done in a discontinuous or continuous way. These two approaches respond to different purposes of biomass valorization and very different costs.

2.3.1. *Production in discontinuous mode*

This method of cultivation is also known as batch culture.

In this system, the algal inoculum and the nutrients required for biomass development are initially introduced into a parent translocation pocket (see Figure 2.9) or into a transparent cylinder with scobalite walls (see Figure 2.10). No addition or removal is made during the cultivation phase. This greatly reduces the risk of contamination of the culture by microorganisms foreign to the cultured species. This mode of production is often recommended when trying to develop a culture in axenic or near-axenic conditions (Becera-Celis 2009).

Closed batch culture is generally based on four stages. The first is the maintenance of the microalgae strain that will be used to produce the mother culture. The strain is maintained and transplanted from a small volume of approximately 250 mL. At this stage, the temperature is between 4 and 22°C (see Figure 2.11). This strain is then cultured in a volume ranging from 250 mL to 4 L and at a temperature ranging from 18 to 22°C. The duration of this culture varies from 7 to 14 days. The obtained mother culture is then transferred to a culture medium with a volume between 4 and 20 L; the temperature and the culture time are identical to those previously mentioned. The resulting intermediate culture is then used as an algal inoculum and can be transferred to a bag or transparent cylinder with a standard volume of 300 L (see Figure 2.10). The species *Haslea ostrearia*, the diatom responsible for the greening of oysters (see section 2.2.3), can be produced in this way, either in a bag (see Figure 2.9) or in a scobalite cylinder (see Figure 2.10).

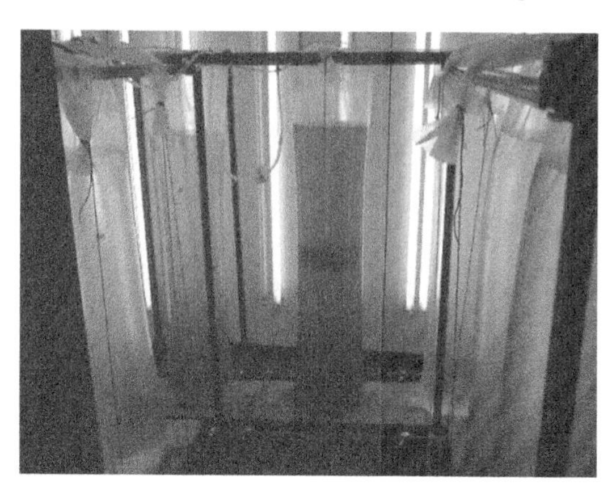

Figure 2.9. *Batch culture of the microalga* Haslea ostrearia *and* Skeletonema costatum *in transparent pockets (photo credit © Rosa P., 2019). For a color version of this figure, see www.iste.co.uk/fleurence/microalgae*

Figure 2.10. *Batch culture in scobalite cylinders of the microalga* Haslea ostrearia *(photo credit © Rosa P., 2019). For a color version of this figure, see www.iste.co.uk/fleurence/microalgae*

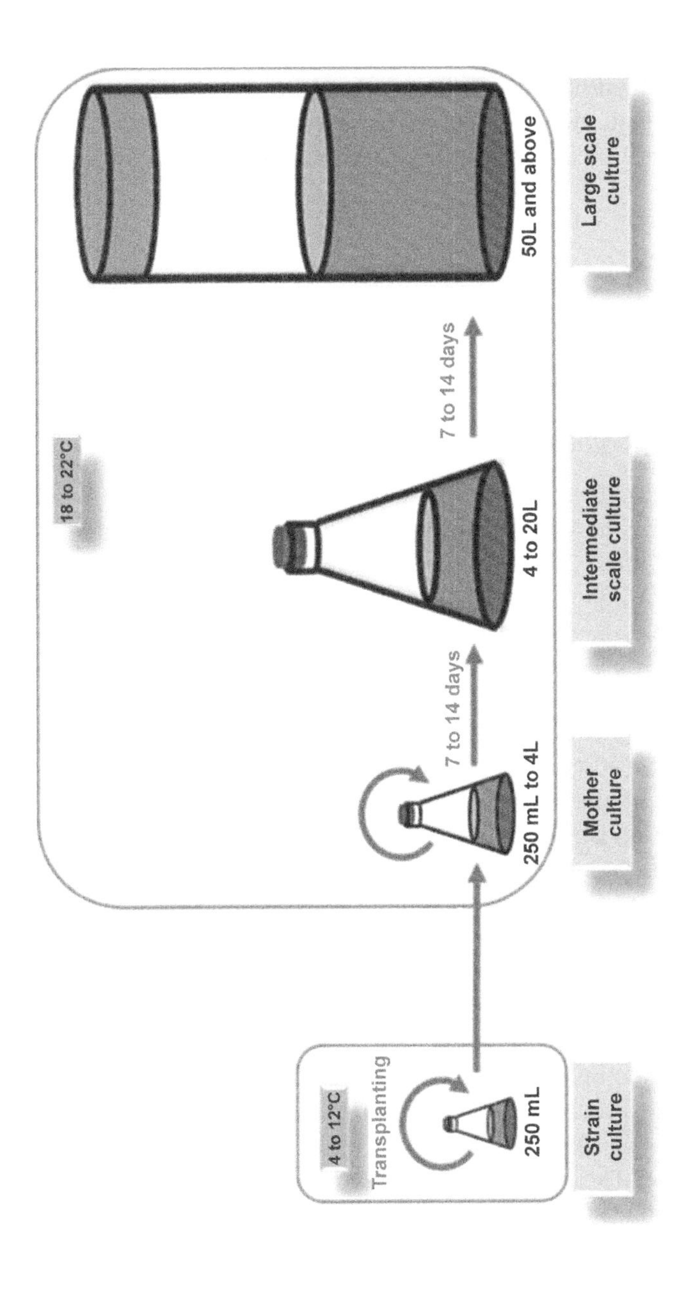

Figure 2.11. *Principle of closed system and batch culture (source: Pouchus Y.-F., based on Helm and Bourne (2006)). For a color version of this figure, see www.iste.co.uk/fleurence/microalgae*

Many species used as fodder algae for shellfish spat are also cultivated in this way. These species are *Skeletonema costatum* or belong to the genera *Isochrysis* and *Tetraselmis* (Helm and Bourne 2006). *Isochrysis galbana* is a species that is frequently cultivated in a closed *batch* system for this type of application. A clone of this species is used as the starting strain for cultivation. This is *Isochrysis* "Tahiti", better known as T-iso by aquaculture professionals.

2.3.2. *Production in continuous mode*

Continuous closed-system production is based on tubular reactors or honeycomb plate technology. These reactors are known as photobioreactors.

2.3.2.1. *Tubular photobioreactors*

There are two main types of tubular photobioreactors, distinguished by the vertical or horizontal nature of the tubes through which the algal culture circulates (see Figures 2.13 and 2.14). The walls of the tubes must allow optimal passage of light and remain transparent throughout the culture. The materials used are mainly glass or plastic derivatives such as methyl polymethacrylate (plexiglass) (Muller-Feuga *et al.* 1998).

In such a system, it is important to maintain agitation of the culture in order to avoid the formation of a biofilm on the walls that could limit the efficiency of the photosynthetic process. A pump injecting air is used to create turbulence and allow light access to all algae in the culture. The pumping generates the circulation of the culture through the tubular network, but this process must not induce the rupture of the cells which have a natural fragility. In order to solve this problem, the frequency of the crop passing through the circuit is reduced and the pumping system adapted. As in the *raceway* system, it is essential to supply CO_2 to the algal biomass during the culture process. Another constraint of this closed system is the evacuation of the oxygen produced by the culture in order to avoid the retro-inhibition of the photosynthetic process by an excess of oxygen in the medium. In a closed system, oxygen thus appears as a limiting factor in the productivity of the photobioreactor. This oxygen is generally eliminated either by bubbling or by passing the culture through a degassing tower. In the latter case, frequent passages of the culture in the degassing tower can end up altering the integrity of the microalgal cell. Culture conditions, such as temperature or light intensity, also influence the level of oxygen produced by the biomass. Thus, for a given light intensity, an increase in temperature is likely to induce an increase in oxygen levels (see Table 2.2) (Torzillo *et al.* 1986).

Light intensity (μE/m^2/s)	Temperature (°C)	Temperature (°C)	Temperature (°C)
	16–20	21–25	26–30
200	0.9	1.3	2.6
400–700	3.0	4.3	5.2
700–1,000	5.0	6.3	7.4

Table 2.2. *Influence of temperature and light intensity on oxygen production (mg O2/L) by a culture of* Spirulina maxima *(*Arthrospira maxima*) in a tubular photobioreactor (from Torzillo* et al. *(1986))*

The temperature rise within the photobioreactor is therefore also a constraint related to the system. This constraint is particularly strong in solar photobioreactors, where the temperature rise increases with the density of the culture. A cooling of the culture via the circulation of a thermally controlled fluid generally remedies this inconvenience. Solar photobioreactors (see Figures 2.13 and 2.14) are not the only types of tubular photobioreactors. There are photobioreactors that operate under artificial lighting. In this type of photobioreactor, light is provided by LED lamps or fluorescent tubes. Typically, the light energy provided by such a system is 1 watt/liter of culture. Artificial light photobioreactors are referred to as ALP (Artificial Light Photobioreactor) (see Figure 2.12) (Muller-Feuga *et al.* 1998). This type of photobioreactor has been compared with other culture systems (e.g. culture in transparent tanks, solar photobioreactors) for the production of the red microalgae *Porphyridium cruentum*. The productivity of such a system was evaluated to be 10 times that reported for a conventional, transparent tank culture system such as that used in hatcheries (Muller-Feuga *et al.* 1998). The cost of the biomass produced by these different systems has been estimated. The cost of biomass is significantly higher (US$2.79/g biomass) in the hatchery system than that reported for natural or artificial light photobioreactors (US$0.74 to US$1.81/g biomass) (see Table 2.3).

Cost	Hatchery tanks	Photobio artificial light reactor (1.5 m²)	Photobio artificial light reactor (10 m²)	Photobio light reactor natural (10 m²)
US dollar(s)/g of biomass	2.79	1.81	0.79	0.74

Table 2.3. *Cost of production of* Porphyridium cruentum *biomass by reactor type (from Muller-Feuga* et al. *(1998))*

Figure 2.12. *Vertical tubular photobioreactor illuminated by LED lamps (photo credit © CEA, 2020). For a color version of this figure, see www.iste.co.uk/fleurence/microalgae*

Figure 2.13. *Horizontal tubular solar photobioreactor growing* Porphyridium cruentum *(photo credit © Braud J.-P., 1990). For a color version of this figure, see www.iste.co.uk/fleurence/microalgae*

Figure 2.14. *Vertical tubular solar photobioreactor growing* Chlamydomonas *sp. (photo credit © CEA, 2020). For a color version of this figure, see www.iste.co.uk/fleurence/microalgae*

Tubular photobioreactors, whether solar or artificial light, are used for the production of microalgae and cyanobacteria. However, this technology needs to be adapted according to the species, as the fragility of the cultured cells differs from one species to another. The resistance of microalgae to the turbulent flow required for optimal light exposure of the culture may explain the differences in productivity between species.

The type of pump used and the nature of the pumping are also determined to ensure cell integrity and culture productivity (see Table 2.4) (Gudin and Chaumont 1991). Positive displacement pumps, whose rotor speed is proportional to the flow rate of the culture, are less damaging to the culture than centrifugal pumps. The latter are, in fact, characterized by a high and constant rotational speed, as well as a longer contact time between the crop and the stator. The presence or absence of a cell wall is also a determining biological factor (see Table 2.4), along with the effect of pumping on the lack of growth of the *Chlamydomonas reinhardtii* mutant.

In addition to this, many parameters brought together under the heading of photobioreactor design also play a role in crop productivity. These parameters are, respectively, the irradiance, the volume and circulation speed of the culture, the diameter of the tubes used, and the total length of the tubular circuit (Gudin and Chaumont 1991; Molina *et al.* 2001). As an example, the culture of the alga *Porphyridium cruentum* was optimized in a solar photobioreactor whose main technical characteristics were to have a culture volume of 6 m^3 and a circuit of 1,500 m of tubes with a diameter of 64 mm (Gudin and Chaumont 1991). Given the climatic conditions, the maximum biomass produced was 25–30 g of biomass per day in summer, compared to 10–15 g of biomass per day in spring and autumn (Gudin and Chaumont 1991). For the culture of *Phaeodactylum tricornutum*, the optimal photobioreactor configuration is obtained for 60 mm diameter tubes, an 80 m tube circuit, a culture circulation speed of 0.5 m/s and an irradiance of 114.64 $\mu E/m^2/s$ (Molina *et al.* 2001).

Types of pumps	*Porphyridium cruentum*	*Chlamydomonas reinhardi*	*Chlamydomonas reinhardi* (wall-less mutant)	*Dunaliella bardawil*	*Haematococcus pluvialis*
Centrifugal pump	+	+	–	+	+
Eccentric pump	++	++	–	++	++
Trilobs pump	+++	+++	–	+++	+++
Screw pump	++++	++++	–	++++	++++
Cellular fragility	Low	Low	High	High	

Table 2.4. *Influence of pump type on the growth (+) and cellular fragility of different species of microalgae grown in tubular solar photobioreactors (from Gudin and Chaumont (1991))*

2.3.2.2. *Planar alveolar photobioreactors*

Planar alveolar photobioreactors are mainly made from plexiglass plates. They are known as VAP (Vertical Alveolar Plates) (see Figure 2.15). These vertically arranged alveolar plates have a thickness that can vary from 1 to 6 cm and a surface area between 0.5 and 2 m^2 (Tredici *et al.* 1991). Other configurations are also possible, such as the one based on a 7 cm thick panel with a surface area of 3.75 m^2 (Sierra *et al.* 2008).

Planar photobioreactors are used for the production of cyanobacteria and microalgae. As for tubular photobioreactors, there are solar reactors and others powered by an artificial light system.

Figure 2.15. *Vertical planar photobioreactor of the Subitec culture type of* Tisochrysis lutea *(photo credit © CEA, 2020). For a color version of this figure, see www.iste.co.uk/fleurence/microalgae*

Sunlight photobioreactors have been successfully used for the production of *Arthrospira platensis* (*Spirulina platensis*) biomass. However, the productivity of the crop depends on the irradiance and thus the incidence of solar radiation. However, this parameter greatly varies according to the seasons. In order to take this criterion into account, vertical plane photobioreactors can be subjected to different inclinations (Tredici *et al.* 1991). During the winter period, the south-facing photobioreactor can be subjected to degrees of inclination with respect to the horizontal, ranging, respectively, from 63°, 67° and 57° for the months of November, December and February. In this winter context, the best productivity of *Spirulina* biomass is obtained in February with a 2.2 m^2 photobioreactor with a 57° inclination (see Table 2.5).

In another seasonal context, the same type of photobioreactor but with a smaller surface area (0.5 m^2) with a 25° inclination to the horizontal generates a productivity of 23.9 g/m^2/day of *Spirulina platensis* biomass. This type of productivity is observed in the month of September with solar radiation that is almost double that recorded during the months of November and February (Tredici *et al.* 1991).

Month	Solar radiation (MJ/m^2/d)	Degree of inclination of the reactor	Productivity (g/m^2/d)
November	8.4	63 °	13.6
December	5.7	67 °	11.8
February	8.9	57 °	14.9

Table 2.5. *Evolution of biomass productivity of* Arthrospira platensis *(*Spirulina platensis*) in a planar alveolar photobioreactor as a function of the winter months, solar radiation and degree of inclination of the reactor (from Tredici* et al. *(1991))*

As with tubular photobioreactors, the designs of this type of equipment, i.e. the surface area used or the degree of inclination provided, are parameters influencing the production of biomass.

The production of microalgae and cyanobacteria is based on two technologies known as open or closed systems. For each technology, the production mode can be either discontinuous or continuous.

This cultivation activity is at the origin of the microalgae economic value chain.

The production of microalgae for commercial purposes began in the 1960s in Japan. It was mainly the cultivation of *Chlorella*, which was used as a dietary supplement (Brennan and Owende 2010). Global production is currently more diversified and includes the cultivation or harvesting of cyanobacteria such as *Spirulina* (*Arthrospira* sp.) or *Aphanizomenon* sp. However, the main mode of production remains the cultivation in *raceway*-type tanks, given the low cost of this type of installation. Countries such as China, India, the United States, Japan and Burma are among the main players in this global production (see Table 2.6). The price of the biomass produced greatly varies depending on the type of application. It ranges from 36 Euros/kg for human food applications to prices of several thousand Euros/kg when this biomass is used as an extraction source for high value-added pigments (e.g. Astaxanthin) (see Table 2.6).

Algae	Annual world production	Main producing countries	Type of application	Biomass price
Spirulina (Arthrospira)	3,000 tons	– China – India – United States – Burma – Japan	Human nutrition	36 €/kg
			– Animal nutrition – Cosmetics	N/A
			Phycocyanin	11 €/mg
Chlorella	2,000 tons	– Taiwan – Germany – Japan	Human nutrition	36 €/kg
			Cosmetics	N/A
			Aquaculture	50 €/L
Dunaliella salina	1,200 tons	– Australia – Israel	Human nutrition	36 €/kg
			Cosmetics	N/A
			β-carotene	2,150 €/kg
Aphanizomenon	500 tons	– United States	Human nutrition	N/A
Haematococcus pluvialis	300 tons	– United States – India – Israel	Aquaculture	50 €/L
			Asthaxanthin	7 150 €/kg

Table 2.6. *Global production status of microalgae and cyanobacteria across all modes (based on Brennan and Owende (2010))*

3

Food Valorizations

3.1. Animal feed

Microalgae and cyanobacteria are also used in animal nutrition. They can be used as fodder algae for feeding bivalve mollusks or for other species, such as crustaceans. They are also used in animal nutrition as feed supplements, mainly as a source of protein as well as a source of pigments or polyunsaturated fatty acids.

3.1.1. *Forage microalgae*

Microalgae are at the base of the oceanic food web. They are consumed via filtration by bivalve mollusks and certain organisms in the larval stage, such as shrimp or abalones (Zmora *et al.* 2013). As such, microalgae are foods of choice for the nutrition of mollusk, echinoderm and shrimp larvae.

3.1.1.1. *Forage microalgae in shellfish nutrition*

Mollusk aquaculture is a very important activity worldwide, showing a strong increase (+30%) over the decade 2001–2010 (FAO 2012). The main species produced are oysters, scallops, clams and mussels, respectively (Muller-Feuga 2000). China, with a production of 11.3 million tons, represents 80% of the world aquaculture activity for filter-feeding mollusks (Muller-Feuga 2013). However, all this activity requires the use of microalgae as fodder for the nutrition of animals in the larval and post-larval stages. In Asia, the main species concerned by this activity are, respectively, *Crassostrea plicatula*, known as the Chinese oyster, and *C. rivularis*, an oyster cultivated in Japan under the name Suminoe oyster. In Europe and more particularly in France, the species bred is *Crassostrea gigas*. The feeding of 10^6 larvae of this mollusk in hatcheries requires the supply of a microalgal biomass of 12 g (expressed in dry matter). The same number of individuals passing to the juvenile stage (0.2–0.3 mm) requires the supply of a biomass of 200 g (Muller-Feuga 2013). The number of species used as forage algae for larval and juvenile feeding is, however, limited to about 10 species (Robert and Trintignac 1997). Of these, eight represent 90% of the algal biomass grown in hatcheries for feeding bivalve mollusk larvae and spat (Robert and Trintignac 1997). The species *Isochrysis affinis galbana* "Tahiti", better known as T-iso, *Chaetoceros gracilis* and *Tetraselmis suecica* are among the most frequently used species (see Table 3.1) (Muller-Feuga 2013).

Species of microalgae	Frequency of use
T-iso	72%
Chaetoceros gracilis	53%
Chaetoceros calcitrans	37%
Tetraselmis suecica	35%
Thalassiosira pseudonana	33%
Pavlova lutheri	26%
Isochrysis galbana	19%
Skeletonema costatum	14%
Dunaliella sp.	9%

Table 3.1. *Examples of microalgae species used in hatcheries for feeding filter-feeding mollusk larvae and juveniles (from Muller-Feuga (2013))*

Mollusk hatcheries and nurseries use live microalgae as forage algae. This requires synchronizing the production of the algal biomass with that of the mollusks. This synchronization is not easy to achieve and, in order to avoid a phytoplankton deficit, the algal biomass produced is often in excess of food requirements of larvae and juveniles. This situation generates an additional cost of the activity, and phytoplankton production can represent up to 30% of the cost of operating hatcheries (Coutteau and Sorgeloos 1992). Since fresh phytoplankton production is considered an economic limiting factor, other possibilities have been considered. These include the use of artificial diets based on dehydrated microalgae, seaweed pastes, microencapsulated products or feeds based on yeasts (Coutteau and Sorgeloos 1992). The diet based on dehydrated yeast, however, shows a low dietary value for juvenile *C. gigas*, when the latter enters into the composition of 25% of the total *C. gigas* diet. 100% of the plan provided. Similarly, the diet based on microencapsulated products induces poor growth on oyster spat (Coutteau and Sorgeloos 1992). Diet based on seaweed paste is found to be more effective, especially as a supplement to the diet based on fresh algae. It can effectively replace 50% of the fresh algal biomass in the diet of *C. virginica* oyster brood stock and spat. The same observation is made for dehydrated seaweed and, in particular, for the species *Tetraselmis suecica*, which is frequently used as a fresh fodder seaweed (see Table 3.1).

In addition to dehydration or formulation processes, another approach based on the conservation of fresh algal biomass at low temperatures was also evaluated. This approach requires the use of microalgae that can withstand concentration and conservation processes at low temperatures. Prasinophyceae belonging to the genus *Tetraselmis* (see Table 1.2) respond to these specificities. Two species, *T. striata* and *T. chui*, were tested in the fresh state to determine which had the best nutritional value for the development of larvae and juveniles of the *C. gigas* oyster. These two species used as the sole food source did not show sufficient nutritional value to promote larval development. In spite of this, the use of *T. striata*, in concentrated form and preserved at low temperature, makes it possible to substitute in a bispecific diet 50% of a microalga with high nutritional potential such as *Chaetoceros calcitrans* without negative impact on larval growth (Ponis *et al.* 2002). The complete substitution of the fresh algal biomass by a transformed biomass or by derived products thus appears to be a solution that is not totally adapted to the zootechnical constraints inherent to the development of larvae and the growth of oyster spat. This has been observed for other species of bivalve mollusks such as clams (*Mercenaria mercenaria*) or scallops (*Patinopecten yessoensis*) (Coutteau and Sorgeloos 1992).

In addition to bivalve mollusks, aquaculture of other species such as abalone (*Haliotis* sp.) is a growing activity. Haliculture is particularly developed in China, Korea, Chile and Australia (Garcia Bueno 2015). China, with an aquaculture production of 56,000 tons, is the world's largest producer of abalone. Korea and South Africa, with a production of 6,228 and 1,015 tons, respectively, follow China's world leader (Garcia Bueno 2015). As with other mollusks, abalone farming relies on hatcheries for larval development. Abalone larvae nutrition can use benthic microalgae present in the form of biofilms deposited on plexiglass plates in larval development tanks (personal communication, Turpin V.). These microalgae of the genera *Navicula* and *Nitzschia* will serve as a diet for abalone juveniles that have a grazer-type feeding mode and facilitate the fixation of the juveniles on a culture medium (Zmora *et al.* 2013). In other nurseries, larvae are fed by injecting seawater naturally rich in phytoplankton into the rearing tanks. However, in the latter case, the microalgae composition and the nutritional value provided are often unknown.

Apart from the species previously mentioned, the diatom *Cocconeis duplex* and the Prasinophyceae *Tetraselmis suecica* are also used for larval feeding of abalone (Becker 2013a).

3.1.1.2. *Forage microalgae in crustacean nutrition*

In 2016, marine and coastal aquaculture produced nearly 4.9 million tons of crustaceans worldwide (FAO 2018a), mainly shrimp. According to OECD projections, its aquaculture production is expected to grow by nearly 35% over the decade 2018–2027 (FAO 2018b). As with bivalve mollusks, the nutrition of shrimp in the larval stage involves the consumption of microalgae. The species used belong mainly to the diatom group and more secondarily to the green or brown algae group (see Table 3.2) (Zmora *et al.* 2013). The contribution of certain diatoms, such as *Chaetoceros calcitrans*, or green algae, such as *Tetraselmis suecica*, in the diet of Japanese shrimp (*Penaeus japonicus*) significantly improves the growth and survival rates of the larvae (Muller-Feuga 2013). However, this finding cannot be generalized to all shrimp species. This is particularly the case for the giant tiger shrimp (*Penaeus monodon*), where the larval survival rate differs according to the microalgae used as a basis for the diet. The use of the species *Dunaliella tertiolecta* in the formulation of the diet leads, in particular, to a low survival rate (23.6%) of the larvae of this crustacean (Kurmaly *et al.* 1989) (see Table 3.3).

Apart from microalgae, cyanobacteria are also used as feed in the nutrition of shrimp larvae. Spirulina (*Arthrospira platensis*) is used in particular in the larval nutrition of the penaeid shrimp (*Penaeus japonicus*) and the brown shrimp (*Penaeus subtilis*) (Becker 2013a). The size of the forage algae used in the larval feeding of bivalve mollusks and shrimp is between 2 and 20 μm. However, the size can reach 100 μm when Spirulina is used as a food source (Zmora *et al.* 2013). The rearing of other crustaceans, such as crabs, also uses microalgae for larval nutrition. Mariculture of crabs is an industrial activity in Asian countries. China, with a production of nearly 800,000 tons, is the world's leading player in this aquaculture activity. This production mainly concerns the species *Eriocheir sinensis*, known as the Chinese crab. In hatcheries, the larvae (zoé) are stored in indoor tanks and fed until the post-larval stage (megalope) with a diet composed of zooplankton, Artemia, yolk and microalgae. The species used belong mainly to the genera *Chaetoceros* and *Chlorella* (Zmora *et al.* 2013). In Japan, the microalgae used are mainly *Nannochloropsis* sp., *Chaetoceros* sp., *Tetraselmis* sp. and *Thalassiosira* sp.

Species	Group
Chaetoceros neogracile	Diatom
Chaetoceros calcitrans	Diatom
Chaetoceros muelleri	Diatom
Skeletonema costatum	Diatom
Thalassiosira weissflogii	Diatom
Thalassiosira fluviatilis	Diatom
Tetraselmis suecica	Green algae
Tetraselmis chuii	Green algae
Isochrysis galbana (T-iso)	Yellow-brown algae

Table 3.2. *Main species of microalgae used as fodder algae for feeding shrimp larvae (based on Becker (2013a), Muller-Feuga (2013) and Zmora* et al. *(2013))*

Species	Survival rate
Rhodomonas baltica	66.2%
Tetraselmis chuii	66.1%
Skeletonema costatum	61.6%
Dunaliella tertiolecta	23.6%

Table 3.3. *Survival rate of larvae of the giant tiger prawn (*Penaeus monodon*) according to the species of microalgae used in the larval diet (from Kurmaly* et al. *(1989))*

In parallel to clear water aquaculture, there is also aquaculture in "green water", in which microalgae are not mainly used as a food source for the nutrition of shellfish larvae or fish, but as oxygenation and purification agents for the rearing environment. This approach, known as *green water* treatment, is a technique often used in hatcheries. Initially, the purpose of introducing microalgae in this technique is to maintain the physico-chemical quality (pH, O_2 level) of the environment in which the larvae develop. In addition to this, the introduction of microalgae into larval rearing tanks is often described as improving the survival, growth and feeding of larvae of many marine species (Haché *et al.* 2017). However, it would appear that the benefits of green water treatment differ depending on the species involved. In the case of penaeid shrimp farming, microalgae seem to mainly play a useful biological filtering role to purify the environment of nitric derivatives toxic to the larvae (Zmora *et al.* 2013). The implementation of green water treatment, based on the use of a mixture of two species *Chaetoceros muelleri* and *Isochrysis galbana*, on American lobster (*Homarus americanus*) larvae does not seem to have any significant effect on their development from the early to the post-larval stage and on their survival (Haché *et al.* 2017). For this species of crustaceans, green water treatment does not also appear to influence larval behavior and food intake (Haché *et al.* 2017). The finding will be somewhat different with fish larvae reared by aquaculture (see section 3.1.1.3).

3.1.1.3. *Microalgae in fish nutrition*

Fish raised by aquaculture are generally carnivorous. Their larvae feed on small prey called rotifers that consume live microalgae. The main species of rotifers used as prey belong to the genus *Brachionus*. The algal species used as a food source for these rotifers are mainly green algae belonging to the genera *Chlorella*, *Dunaliella*, *Nannochloris* and *Haematococcus* (Muller-Feuga 2013). Some species belonging to other groups are also used, such as *Isochrysis galbana* (see Table 3.4). Artificial yeast feeds can also be used to replace microalgae. However, microalgae, due to their biochemical composition (see Table 3.5), have the advantage of improving the nutritional value of the rotifers consumed by the larvae. Their presence also seems to limit the contamination of farms by bacteria of the genus *Vibrio*, a pathogen much feared in aquaculture production (Muller-Feuga 2013).

In addition to their use as fodder algae for rotifer feeding, the species mentioned can also be used to generate green water treatment in larval rearing ponds. This method of aquaculture is described as having many advantages for larval fish farming. Indeed, this technique would improve the growth of fish larvae during their feeding period which is based on the consumption of rotifers (Zmora *et al.* 2013). Other effects, such as increased larval survival rate or maintenance of the sanitary quality of larval rearing, are also observed.

This situation seems to be related to the improvement of water quality inherent to the green water process as well as the production by the microalgae of antibacterial exudates and stimulators of a non-specific immune response of the larvae. Beyond these aspects, the green water treatment would improve the food intake by the fish larvae as well as the nutritional quality of the latter. This last phenomenon seems to be linked to the consumption of rotifers whose nutritional value has been enriched by their supply of microalgae. It can also be explained by the direct ingestion and intestinal absorption by fish larvae of the microalgae present in the rearing basin (Zmora *et al.* 2013).

Class	Genus	Species
Chlorophyceae (green algae)	*Chlorella*	*C. minutissima*
	Dunaliella	*D. virginica*
	Nannochloris	– *N. grosi* – *N. tertiolecta*
	Haematococcus	– *H. salina* – *H. pluvialis*
Prasinophyceae (green algae)	*Tetraselmis*	– *T. suecica* – *T. chuii* – *T. striata*
Prymnesiophyceae (Haptophyte)	*Isochrysis*	*I. galbana "Tahiti"* (T-iso)
	Pavlova	*P. salina*

Table 3.4. *Main species of microalgae that can have a dual use: live prey as forage algae and/or for the production of green water in the rearing of larval fish (from Muller-Feuga (2013))*

The cyanobacterium, *Arthrospira platensis*, better known as Spirulina, can also be used in particular in the nutrition of herbivorous fish. This species, easily available because it is widely cultivated globally, is recognized as an interesting source of protein (see Table 3.5) in animal or human food. The use of Spirulina in the diet of blue tilapia (*Tilapia aurea*) at a daily intake of 29 g (expressed as dry matter) per body kilogram would lead to a daily growth rate of 14.4 g/kg in the animal (Stanley and Jones 1976). Apart from the contribution of Spirulina to growth, its use in the diet of fish may also have a health benefit.

Species	Proteins	Carbohydrates	Lipids
Dunaliella tertiolecta	20.0%	12.2%	15.0%
Chlorella vulgaris	48.0%	8.0%	13.0%
Tetraselmis chui	31.0%	12.1%	17.0%
Tetraselmis suecica	31.0%	12.0%	10.0%
Isochrysis galbana	29.0%	12.9%	23.0%
Spirulina platensis	64.0%	25.0%	7.0%

Table 3.5. *Examples of the biochemical composition of a few species of microalgae and a cyanobacterium used in larval fish farming (values expressed as % of dry matter) (from Becker (2013a))*

This aspect has been particularly described in Nile tilapia (*Oreochromis niloticus*). This species has been the subject of close aquaculture production of 4.2 million tons in 2016 (FAO 2018a) and represents an important source of animal protein in many countries in Africa and Asia. However, Nile tilapia is susceptible to many bacterial pathogens, particularly *Aeromonas hydrophila*, which causes severe septicemia in fish. The incorporation of Spirulina in the diet of tilapia results in a modification of blood parameters such as the number of red blood cells (+26%) and white blood cells (+22%) and is characterized by a significant increase in the number of lymphocytes (+12%) (Abdel-Tawwab and Ahmad 2009). For a preparation incorporating 10 g of Spirulina per kilogram of diet, tilapia mortality is reduced to just under 10% in the presence of *Aeromonas hydrophila* (Abdel-Tawwab and Ahmad 2009). In the absence of Spirulina, this mortality is reported at 80% only four days after contact with the pathogen.

Independently of their use as fodder algae in the direct or indirect feeding of aquatic animals, microalgae, by their richness in certain constituents such as proteins, pigments, minerals, vitamins and fatty acids, can be part of the diet of many livestock. In this case, they are considered to be dietary supplements.

3.1.2. *Dietary supplements*

3.1.2.1. *Poultry feeding*

In particular, trials have been carried out in Israel on the use of microalgae to replace the protein fraction provided by soybean in poultry feed. In these tests, the replacement of 25% of the protein intake from soya by 5% microalgae gave positive results from a zootechnical point of view. On the other hand, the use of 15% microalgae feed ration resulted in a very significant decrease in the efficiency of dietary nitrogen conversion (Becker 2013b). However, these experiments carried out using microalgae developed in wastewater represent an interesting approach in terms of valorization and development of an eco-responsible sector. Other more targeted experiments in terms of species used were also developed in the 1980s. Chlorella biomass, previously transformed by drying at low temperature and grinding in order to improve its digestibility, was introduced into the diet of hens for 16 or 20% of the ration. Such a high-protein diet (71% total protein in the feed ration) did not generate any toxicity in the animals subjected to this diet (Becker 2013b). The use of Spirulina as a feed supplement in poultry feed has also been the subject of much work with often mixed results. In turkeys, the use of Spirulina at doses of 1–10 g/kg of feed ration significantly increases the growth rate of the animals and lowers their mortality rate. The mortality rate is reduced by 12% on the basis of a diet including a content of 1 g of Spirulina per kilogram of feed ration (Becker 2013b). Feed rations of up to 10% Spirulina can be used instead of conventional protein without any negative influence on the health status of the animals. On the other hand, higher concentrations seem to have an opposite impact on the animals, especially in longer diets (Becker 2013b).

The integration of dehydrated Spirulina in the composition of the feed ration of rooster chicks and its influence on their growth have also been evaluated (Ross and Domini 1990). The use of doses ranging from 10 to 20% of Spirulina in the feed ration of 3-week-old chicks revealed a slight negative effect on their growth, compared to that reported for chicks having been subjected to the control diet without Spirulina. This is rather paradoxical, given that Spirulina at this concentration level does not seem to have a negative effect on the feed conversion rate of the animals (see Table 3.6).

Spirulina content	Body growth gain	Food conversion rate
0%	173 g	1.9
5%	167 g	2.0
10%	164 g	2.1
20%	146 g	2.0

Table 3.6. *Effect of Spirulina content in the diet of rooster chicks on their growth and conversion efficiency protein food (based on Ross and Domini (1990))*

Other studies of substitution of traditional protein sources (fishmeal, peanut cake) by dehydrated Spirulina in broiler nutrition have shown more encouraging results (Venkataraman *et al.* 1994). Over a period of 12 days and without the addition of additional vitamins or minerals, a diet consisting of 140–170 g/kg of Spirulina feed does not affect the zootechnical performance of the animals, particularly in terms of body mass, feed intake efficiency and protein conversion rate. The nutritional quality of the flesh remains unchanged, and only the color of the flesh varies compared to that of animals fed without Spirulina. This color is more intense for animals fed a diet containing cyanobacteria. The presence in Spirulina of a wide variety of carotenoid pigments, such as β-carotene, xanthophylls, zeaxanthin or canthaxanthin (see Table 3.7) (Becker 2013b), is probably the cause of the skin coloring phenomenon in chickens. Such an effect on the skin coloration of broiler chickens has also been described from feeding experiments with the microalgae *Haematococcus pluvialis*. In the latter case, the orange coloration of the tissues increases with the amount of microalgae powder included in the feed ration (350–8,950 mg algae/kg feed) (Waldenstedt *et al.* 2003).

In addition to skin coloring, the effect of microalgae supplementation on egg yolk coloring in laying hens has also been investigated (Bruneell *et al.* 2013). The intake of the microalga *Nannochloropsis gaditana* in the feed ration of laying hens in the form of a powder, up to 5% of the feed, results in a bright orange egg yolk, which implies an efficient transfer of the carotenoids contained in the algae. This experiment also showed that the algae-based feed significantly improved the polyunsaturated fatty acid content of the eggs, particularly DHA (docosahexaenoic acid).

The use of microalgae or Spirulina as a source of protein to replace conventional protein sources (soybean, peanut, fish meal) in poultry feed produces contrasting results depending on the species (broilers, layers, chicks), the composition of the basic diet and the time of administration.

Beyond the nutritional aspects inherent to the supply of proteins or pigments for bird feed, the use of Spirulina in chicken feed has also been described as having a beneficial effect on the immune system of the animals (Becker 2013b).

In relation to these results and in view of the production costs of microalgae and Spirulina, the use of this biomass in poultry feed is an interesting prospect, but it does not always correspond to an agro-industrial reality.

3.1.2.2. *Ruminant feeding*

The use of microalgae as feed supplements in ruminant nutrition has been the subject of some experimentation. This type of application has been very little developed because it requires very large quantities of algal biomass and imposes high technical constraints (Becker 2013b).

The first experiments involved feeding the calves a daily supply of 1 L of *Scenedesmus obliquus* solution (3.10^8 cells/mL) over a period of three weeks. The influence of this intake on the digestibility of the feed appears to be very small. On the other hand, this type of feed improves the role of the intestine in the digestion process.

Other tests on lambs, ewes and cattle were performed with a mixture of *Chlorella* sp., *Scenedesmus obliquus* and *S. quadricauda* (Becker 2013b). A content of 60% algal mixture in the animal feed ration was tested. Such a diet led to a decrease in the digestibility of the dietary protein fraction provided, probably due to the high fiber content of the microalgae.

Numerous studies have been carried out on the effect of a diet integrating Spirulina in the milk production of cows, on growth performance, protein digestibility and on the serum parameters of calves.

According to Becker (2013b), this work and particularly that on Spirulina need to be consolidated by other research in order to establish with greater certainty the effect of a microalgae-based diet on ruminant nutrition.

3.1.2.3. *Feeding of pigs*

Little work has been done on the use of microalgae or Spirulina in pig nutrition. The introduction of a mixture of microalgae (*Chlorella* sp. and *Scenedesmus* sp.) grown in wastewater to replace the protein fraction provided by soya in the feed of about 50 pigs was tested.

The evaluated diets contained 2.5%, 5% and 10% algal mixtures in the feed ration provided to the pigs. After a seven-week period, the feed conversion rate varied between 3.76 and 3.90 depending on the algal diet. The latter is very close (3.85) to that observed for animals fed a conventional control diet (Becker 2013b).

This same constant is observed in the growth of pigs. The provision of a Scenedesmus-based diet to replace the protein fraction provided by fish meal or soybean meal does not produce a significant difference in the growth of animals on such a diet compared to that reported for pigs fed a conventional diet.

The same type of experimentation was carried out with two cyanobacteria *Spirulina maxima* and *Spirulina platensis* (*Arthrospira platensis*), and a microalga (*Chlorella* sp.). This mixture was used to replace 33% of the protein fraction provided by soy in the traditional diet. No effects, such as loss of appetite, intestinal disorders or damage to the gastrointestinal tract of animals weaned from the algal diet, were observed. Animals fed this mixture showed an improvement in weight gain compared to control animals. Animals fed with 1 g of algal mixture per kilogram of feed ration showed an improved feed conversion rate (Becker 2013b).

This experiment shows that the mixture of cyanobacteria and microalgae is a nutritionally acceptable substitute for the soy protein fraction with no apparent toxicity to animals.

It is likely that such a mixture can substitute for protein rations based on soybean or fish meal within substitution limits to be determined according to the nature of the algal mixture.

3.1.2.4. *The feeding of rabbits*

Some very rare tests of application of diets based on Spirulina (*Arthrospira platensis*) have also been carried out for rabbit nutrition. The effect of such diets on the animal carcass, meat quality and fatty acid composition of the flesh has been particularly studied (Peiretti and Meineri 2008). This study involved about 40 individuals separated into four groups of 10 (five males, five females). Three diets incorporating Spirulina contents of 5, 10 and 15% in the food ration were administered to them over a period of 24 days. No mortality was observed during the experiment, confirming the absence of acute- or medium-term oral toxicity of the addition of Spirulina in rodent nutrition. No significant differences in weight gain or feed conversion efficiency were observed between the different groups and the control group not fed a Spirulina-based diet. Only the group on a diet containing 10% Spirulina showed a higher level of food intake than the other groups. On the other hand, significant differences were reported in the digestibility of organic matter, protein and crude fiber, which was higher for the control group (diet without Spirulina).

Flesh composition remains unchanged except for the lipid content which is lower in the control group than in the animals on Spirulina diets (Becker 2013b).

This experiment shows that the substitution of soybean or clover in the rabbit feed ration is possible, as it has no negative effects on the survival and growth of the animals. The quality of the flesh, apart from the lipid content, also appears to be unchanged by feeding Spirulina.

The economic cost of such a substitution is, however, a major issue that, to date, does not seem to be very well determined.

3.1.2.5. *The feeding of fish*

As described above, microalgae can be used directly or indirectly as feed in the nutrition of larval fish (see section 3.1.1.3).

There is also another field of application of microalgae that concerns the contribution of this biomass as a source of pigments to accentuate the coloration of the flesh, especially of salmonids. The effects of the supply of a diet based on Chlorella (*Chlorella vulgaris)* on the flesh color of rainbow trout (*Oncorhynchus mykiss*) have been tested (Gouveia *et al.* 1996).

Chlorella has the particularity of having a very varied and rich carotenoid composition (2 g/kg dry matter) (see Table 3.7) (Gouveia *et al.* 1996; Becker 2013b; Morançais *et al.* 2018), which justifies its use as a source of orange-yellow pigments.

Chlorella is included in the feed ration in dehydrated or micronized form up to 4%. The purpose of micronization is to break the wall of microalgae and to promote the bioavailability of pigments. Parallel to the supply of Chlorella, other groups are fed a diet that does not combine microalgae but contains synthetic pigments, in this case astaxanthin and canthaxanthin. After three weeks of feeding, no significant differences in the flesh color of the animals fed the different diets are visually observable when applying the Roche color chart, which is used to assess the flesh color of salmonids. After nine weeks, a very slight increase in color is recorded for fish that have had a synthetic astaxanthin intake in their diet.

The incorporation of microalgae of the genus *Chlorella* as a source of natural pigments therefore gives results comparable to those obtained with synthetic pigments. Pre-treatment of microalgae by micronization does not bring any significant improvement in the deposition of carotenoids in the fish muscle.

This study shows that the use of a raw Chlorella biomass up to 40 g/kg of feed ration is an alternative to the supply of synthetic pigments.

Chlorella **sp.**	***Spirulina*** **sp.**
Astaxanthin	Xanthophyll
β-carotene	β-carotene
Zeaxanthin	Zeaxanthin
Canthaxanthin	Canthaxanthin
Lutein	Lutein
Violaxanthin	Violaxanthin
Neoxanthin	Echinenone
Auroxanthin	Cryptoxanthin

Table 3.7. *Examples of some carotenoid-like pigments present in* Chlorella *sp. and* Spirulina *sp. (from Gouveia* et al. *(1996), Becker (2013b) and Morançais* et al. *(2018))*

3.2. Human food

3.2.1. *Ingredients or vegetables*

Before the 20th century, Spirulina (*Arthrospira platensis*) and *Nostoc* sp. were the only cyanobacteria used directly in human nutrition as an ingredient in their own right (Borowitzka 2018b).

The use of Spirulina by the Aztecs and the people of Lake Chad was justified in view of the protein intake (see Table 3.8) that this organism provides in human food. This cyanobacterium, once dried, was used by the Aztecs to make a bread called "tecuitlatl". For the people living along the shores of Lake Chad, Spirulina, after drying in the sun (see Figure 3.1), is incorporated into the flour used to make wafers or cookies called "dihé" (see Figure 3.2). In this way, people consume 10–12 g of Spirulina per day (Becker 2013b). In the Kanem region (Chad), Spirulina (e.g. *Oscillatoria platensis*) is mainly used to make sauces (dié) with corn or millet balls.

Species	Proteins (% of MS)	Carbohydrates (% of MS)	Lipids (% of MS)
Spirulina platensis	55–70	8–19	4–8
Chlorella sp.	38–58	12–26	2–22
Dunaliella sp.	9–23	4–40	8–25
Haematococcus pluvialis	10–48	27–33	15–40
Isochrysis galbana	27–39	13–17	17–24
Porphyridium cruentum	28–39	40–57	9–14
Soy	37	30	20
Egg	47	4	41

Table 3.8. *Examples of the biochemical composition of Spirulina* (Arthrospira platensis)*, some microalgae and reference foods (from Batista* et al. *(2013) and Becker (2013b))*

Figure 3.1. *Drying Spirulina in the open air and in the sun after harvest (photo credit © FAO/Marzot M., 2010). For a color version of this figure, see www.iste.co.uk/fleurence/microalgae*

Figure 3.2. *Sale of dihé in a Chadian market (photo credit © FAO/Marzot M.). For a color version of this figure, see www.iste.co.uk/fleurence/microalgae*

In this region, Spirulina is used for the preparation of sauces, except for pregnant women who consume it directly in dried form. This particular practice is due to beliefs that are related to witchcraft. Consumption of this cyanobacterium in raw form would darken the womb and prevent the witch doctor from seeing the child according to local beliefs (Delpeuch *et al.* 1976). In the arid areas of western China, the cyanobacterium *Nostoc flagelliforme* is harvested for the preparation of soups traditionally consumed during Chinese New Year celebrations. This bacterium, marketed under the names *fa cai* in Mandarin and *fat choy* in Cantonese, is a valuable ingredient that sells for approximately 110 euros per kilogram (Roney *et al.* 2009). The use of diatoms for the production of ingredients has also been tested in wartime. Harder and von Witsch (1942) mass-grew the species *Nitzschia palea* for emergency production of dietary fatty acids (Becker 2013b). Although qualified as a "superfood" by NASA, Spirulina and microalgae are used more as an ingredient than as a food product. Spirulina is used, in particular, in the composition of pasta. This type of product generally consists of 97% wheat flour with 3% Spirulina (*Arthrospira platensis*) added (source Bjorg site). Concerning Chlorella, it is the main microalgae consumed in human food, either as an ingredient or as a food supplement (see section 3.2.2). Its use as an ingredient has been described in particular for the elaboration of "plankton bread" (see Figure 3.3). This bread combines Spirulina and Chlorella as ingredients. In addition, many suppliers of Chlorella powder offer original culinary preparations in which microalgae is used as an ingredient. Among the proposed recipes, we can mention the

“Forest crunchy”, the “Chlorella sea terrine” or the “Chia, Chlorella and strawberry pudding” (source: eChlorial).

The microalgae *Odontella aurita* is also used in the composition of a vegan food, “Solmon”, which is made up of a macerate of microalgae and seaweed. This reconstituted product is marketed in smoked vegetable slices and is presented as a substitute for smoked salmon for vegans. In particular, it is used in many preparations as an alternative to salmon (see Figures 3.4 and 3.5).

Figure 3.3. *Plankton bread (photo credit © Cadour P., 2020). For a color version of this figure, see www.iste.co.uk/fleurence/microalgae*

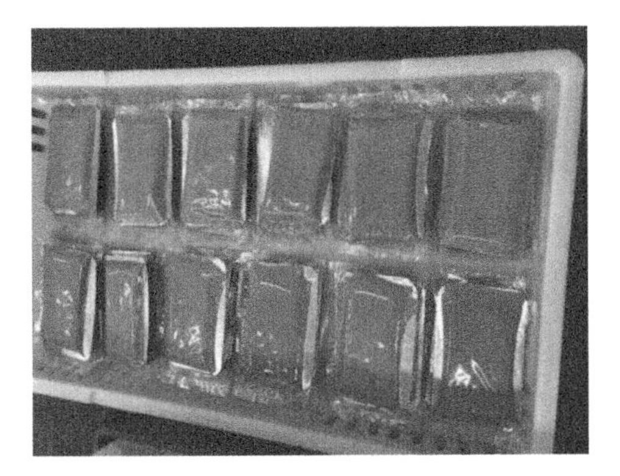

Figure 3.4. *Solmon product incorporating the microalgae* Ondotella aurita *(photo credit © Odontella, 2020). For a color version of this figure, see www.iste.co.uk/fleurence/microalgae*

Figure 3.5. *Culinary preparation based on Solmon (*Ondotella aurita*) (photo credit © Odontella, 2020). For a color version of this figure, see www.iste.co.uk/fleurence/microalgae*

Apart from applications as ingredients, microalgae and cyanobacteria are mainly used in human nutrition as dietary supplements with commercial "health food" type claims.

3.2.2. *Dietary supplements*

Microalgae and cyanobacteria are mainly used in human nutrition in the form of food supplements. These are mainly marketed in the form of tablets, capsules or powder. The main cyanobacteria used are distributed under the generic name of Spirulina and are referred to as algae or even blue-green algae. This name is based on an old classification in which, indeed, Spirulina was listed as belonging to the class of Cyanophyceae (blue algae). Since then, the name of this class has been changed to the class of cyanobacteria. The name "algae" for Spirulina is, therefore, a scientifically usurped name since Spirulina are not eukaryotic organisms but belong to the evolutionary lineage of eubacteria. This scientific reality is rarely taken into account by producers and distributors of food supplements, probably because the term "algae" is more popular than "bacterium". Another confusion concerns the genus name Spirulina. Indeed, *Spirulina platensis*, the Spirulina most frequently produced, has been renamed by taxonomists *Arthrospira platensis* since the late 1980s (Becker 2013b). This also applies to the species *Spirulina maxima*, which is today called *Arthrospira maxima*. This evolution of nomenclature is still largely ignored by consumers, distributors and regulations (see section 3.2.5). The cyanobacterium *Aphanizomenon* sp.

harvested from Lake Klamath (United States) is also marketed as a food supplement under the name "Klamath algae" or "klamath bio". The latter case is quite interesting because the commercial effect is based more on the originality of the geographic site than the organism itself.

Spirulina (*Arthrospira platensis*), used as a dietary supplement, is mainly produced in China. In 2009, production in this country was 3,500 tons (expressed in dry matter), 40% more than three years earlier (Lu *et al.* 2011). This crop is partly grown (20% of production) in the highlands of Inner Mongolia, which benefit from a sunny climate conducive to this kind of activity (Lu *et al.* 2011). The United States and its two production sites, on the coast of California and in the Hawaiian Islands, appear to be the world's second-largest producer of this cyanobacterium, with a volume of nearly 1,000 metric tons (expressed in dry matter). In Europe, France and Hungary are the two main producers of Spirulina for dietary supplements.

Chlorella, a green microalga, is also highly valued as a dietary supplement. Previously, this microalgae was used in folk medicine and recognized as a traditional food in Asia. Commercial production of Chlorella started in the 1960s in Japan, and its production is about 2,000 tons (expressed as dry matter) (see Table 2.6). Currently, it is cultivated for use in food supplements that are often marketed with "health" claims (see section 3.2.3). The main producing countries remain China and Japan, with the world production estimated at 2,000 tonnes/year (expressed on a dry matter basis) (Batista *et al.* 2013). The diatom, *Odontella aurita*, is also valued as a dietary supplement. However, its recent production appears marginal compared to that of Spirulina and Chlorella, and there is little validated data as a dietary supplement. The justification for the use of microalgae and cyanobacteria as food supplements is essentially based on the richness or originality of certain constituents such as minerals, vitamins, polyunsaturated fatty acids or pigments. The mineral content of microalgae or cyanobacteria can vary, as with macroalgae, depending on the culture medium. However, there is enough data to establish average or even indicative mineral compositions. Apart from the influence of the culture medium on the composition, the most important factor is the interspecific variability. For example, the species *Chlorella vulgaris* has been shown to have a calcium content about 5 times higher than that reported for Spirulina and other microalgae (see Table 3.9) (Batista *et al.* 2013). In the same vein, the microalga *Haematococcus pluvialis* has an iron content 9 times higher than that of Spirulina and 5 times higher than that of Chlorella (see Table 3.9). Such variations are also reported for magnesium, an important trace element in human nutrition. Chlorella, with a magnesium content of about 1.5% (expressed on a dry matter basis), appears to be an interesting source in the food supply of this element. However, this content is dependent on growing conditions: the same species under different conditions may show much lower magnesium contents (0.36–0.44%) (Safi *et al.* 2014). In spite of this observation, the designation of superfood given by the Japanese does not seem to be usurped with regard to the mineral composition of this microalga. Spirulina appears interesting in relation to its nitrogen content (7.2% expressed on a dry matter basis) which is higher than that of Chlorella. This nitrogen is mainly in the form of proteins. Spirulina, used as an ingredient or as a dietary supplement, therefore appears more as an interesting source of protein than as a source of minerals, compared to Chlorella. Concerning the vitamin composition, the observation is more nuanced. Indeed, Spirulina shows significantly higher vitamin A, C and E contents than those reported for Chlorella (see Table 3.10) (Becker 2013b). Its vitamin A content is about 840 mg/kg dry matter. This is significantly higher than that of vegetables such as spinach (130 mg/kg dry matter), which is known to be rich in vitamin A.

	Spirulina maxima	*Chlorella vulgaris*	*Haematococcus pluvialis*
N (%)	7.19	6.08	1.64
P (%)	1.29	1.53	1.31
K (%)	2.58	0.98	0.97
Ca (%)	0.91	4.73	0.25
Mg (%)	0.35	1.46	0.22
Fe (mg/kg)	93.60	166.30	822.70
Cu (mg/kg)	1.10	2.20	344.00
Mn (mg/kg)	24.60	471.5	111.90
Zn (mg/kg)	3.50	17.5	232.20

Table 3.9. *Examples of the mineral composition of Spirulina* (Arthrospira maxima*) and some microalgae (from Batista* et al. *(2013))*

Apart from the vitamin aspects, some microalgae and Spirulina have a relatively interesting, even original, profile in polyunsaturated fatty acids and sterols. In this context, they can complement with advantage other dietary sources of fatty acids of nutritional interest, such as those provided by salmonids (salmon, trout).

In the species, *Spirulina maxima* (*Arthrospira maxima*), the fraction of polyunsaturated fatty acids represents 35% of the total fraction of fatty acids. It is characterized in particular by a richness in linolenic acid ɣ (C18:3ω 6) (452 mg/100 g dry algal biomass) (Batista *et al.* 2013) (see Table 3.10). On the other hand, this species seems relatively low in unsaturated fatty acids of series ω3, and more particularly in eicosapentaenoic acid (EPA) and docosahexaenoic acid (DHA) which it seems to lack. On the other hand, the species *Chlorella vulgaris* is mainly characterized by high levels of typeω 3 polyunsaturated fatty acids (972 mg/100 g dry matter). It shows EPA and DHA contents of 19 and 16 mg/100 g dry matter, respectively. However, the species *Haematococcus pluvialis* and *Isochrysis galbana* differ from the previous ones by their levels of EPA and DHA in exceptional omega-3 polyunsaturated fatty acids (5,772–6,462 mg/100 g dry matter). By way of comparison, salmon, depending on its origin, i.e. wild or farmed, contains 2,200–4,700 mg/100 g of flesh (Bourre *et al.* 2006). These two species of microalgae are also distinguished by remarkable EPA contents of 579 and 4,875 mg/100 g dry matter, respectively (see Table 3.11). Farmed salmon fed with fishmeal and fish oils show an EPA content of 1,100 mg/100 g flesh. Wild salmon have a flesh content of 600 mg/100 g flesh. Certain microalgae, therefore, appear to be interesting sources of polyunsaturated fatty acids in a complementary way to other sources of animal origin. This is particularly the case of *H. pluvialis* and *I. galbana*, which are already used as food supplements, even if their use remains more limited than that of Spirulina or Chlorella.

The use of microalgae as food supplements is also based on the presence and diversity of pigments present in this resource. Many pigments are known for their antioxidant power; this characteristic is at the origin of the use of microalgae or Spirulina in human nutrition. Spirulina, a cyanobacterium, has a characteristic pigment, phycocyanin, belonging to the family of phycobiliproteins (see Table 3.12). This original pigment is known for its antioxidant and anti-inflammatory properties (Fleurence and Levine 2018) (see section 3.2.3). Other pigments belonging to the carotenoid family, such as astaxanthin, have been described as having a protective effect against several types of cancers (Nazih and Bard 2018). The antioxidant properties of pigments are a commercial argument often developed for the diffusion of food supplements based on microalgae.

The nature of these pigments varies according to the species of microalgae or cyanobacteria (see Table 3.12). With the exception of chlorophyll and carotenoid pigments, some microalgae have original pigments not found elsewhere. This is particularly the case for diatoms of the genus *Haslea*, which have a very special pigment called marennine. This greenish-blue pigment is responsible for the greening of the oysters that consume these microalgae (see section 2.2.3).

Species	Vitamin A (mg/kg)	Vitamin C (mg/kg)	Vitamin E (mg/kg)	Vitamin B1 (mg/kg)	Vitamin B2 (mg/kg)	Vitamin B6 (mg/kg)
Spirulina platensis	840.0	80.0	120.0	44.0	37.0	3.0
Scenedesmus quadricauda	554.0	396.0	–	11.5	27.0	–
Chlorella pyrenoidosa	480.0	–	–	10.0	36.0	23.0
Recommended daily intake (mg/day)	1.7	50.0	30.0	1.5	2.0	2.5

Table 3.10. *Examples of the vitamin composition of Spirulina (*Spirulina platensis *or* Arthrospira platensis*) and some microalgae (from Becker (2013b)) (the values are determined in relation to the dry matter)*

	Spirulina maxima (mg/100g)	*Chlorella vulgaris* (mg/100g)	*Haemotococcus pluvialis* (mg/100g)	*Isochysis galbana* (mg/100g)
Saturated fatty acids	1,146	1,254	7,722	6,681
Monounsaturated fatty acids	402	836	13,387	4,213
Polyunsaturated fatty acids	58	971	5,570	6,461
C18:3 ω6	452	112	472	–
C20:5 ω3 (EPA)	–	19	579	4,875
C22:6 ω3 (DHA)	–	16	–	1,156

Table 3.11. *Examples of fatty acid composition of Spirulina (*Spirulina maxima *or* Arthrospira maxima*) and some microalgae (from Batista* et al. *(2013)) (the values are determined in relation to the dry matter)*

	Spirulina maxima	*Chlorella vulgaris*	*Haematococcus pluvialis*	*Isochrysis galbana*	*Haslea ostrearia*
Main fat-soluble pigments	– Lutein – Zeaxanthin – β-carotene – Chlorophyll a	– Astaxanthin – Lutein – Canthaxanthin – Zeaxanthin – Chlorophylls a and b	– Astaxanthin – Lutein – Canthaxanthin – Zeaxanthin – Chlorophylls a and b	– Fucoxanthin – Lutein – Zeaxanthin – β-carotene – Chlorophylls a and c	– β-carotene – Fucoxanthin – Diatoxanthin – Diadinoxanthin – Chlorophylls a and c
Main water-soluble pigments	– Phycoerythrin – Phycocyanin – Allophycocyanin	–	–	–	Marennine

Table 3.12. *Examples of the main pigment composition of Spirulina (*Spirulina maxima *or* Arthrospira maxima*) and some microalgae (from Batista (2013) and Morançais* et al. *(2018))*

The use of microalgae or cyanobacteria as food supplements is a well-known application for the general public. The use of this resource as a functional food is more delicate and often controversial, as this notion is dependent on "health" claims that are not always proven and are subject to more or less permissive regulations, depending on the country.

3.2.3. *Functional foods*

The concept of functional food was born in Japan in the 1980s. At that time, it was defined as a food for health use. This definition was taken up in Europe under the term "alicament". This concept was very strongly discussed, especially since the associated claims, even if they referred to prevention, concerned serious pathologies, such as cardiovascular diseases, cancer or certain viral infections.

It seems to be accepted today that a food alone cannot constitute a preventive food medicine. Indeed, only a varied diet appears to be the means of prevention recommended by all nutrition specialists.

This development having been made, it is nevertheless true that many studies associate interesting therapeutic biological activities with microalgae or at least with some of their molecules. These activities have often been determined in *in vitro* cell models and sometimes *in vivo* in animal models. On the other hand, there is little information on the metabolism of these molecules in humans. In spite of this, it is interesting to draw up a balance sheet of these activities of interest for health in order to better understand the epidemiological studies that report a beneficial effect of the integration of algae in the diet.

3.2.3.1. *Antiviral activities*

The antiviral activities associated with certain microalgae and cyanobacteria were initially identified from aqueous or alcoholic (methanol or ethanol) extracts. In particular, certain aqueous extracts of *Spirulina platensis* showed inhibitory activity of the reverse transcriptase of the virus HIV-1 present in cultured human cell lines (Dewi *et al.* 2018). Antiviral activities detected in extracts of microalgae or Spirulina are often related to the presence of polysaccharides, mostly sulfated (see Table 3.13). For example, the sulfated exopolysaccharide of the red microalga *Porphyridium cruentum* develops anti-viral activity against the herpes virus (types 1 and 2) *in vitro*, as well as *in vivo* in rat and rabbit models. In the same vein, the *Navicula directa* diatom also produces a sulfated exopolysaccharide, naviculan, which also shows activity against the herpes virus, as well as against the influenza virus by early blocking of the replication phases of these RNA viruses in host cells. In addition to polysaccharides, certain microalgae and cyanobacteria also produce proteins with antiviral properties. These are mainly glycoproteins that can be related to lectins. This is notably the case of cyanovirin-N produced by the cyanobacterium *Nostoc ellipsorum*, which has strong antiviral activity against the HIV virus and the herpes virus (Dewi *et al.* 2018).

Species	Type of compound	Virus
Porphyridium cruentum	Sulfated polysaccharide	–Herpes 1 and 2 – Vaccine (cowpox)
Navicula directa	Sulfated polysaccharide (naviculan)	– Herpes 1 and 2 – Influenza A, HIV-1
Chlorella vulgaris	Sulfated polysaccharide	Herpes 1
***Arthrospira platensis* (*Spirulina platensis*)**	Sulfated polysaccharide (calcium spirulane)	– Herpes 1 – Influenza – Polio – Cytomegalovirus
Nostoc flagelliforme	Acidic polysaccharide (nostoflan)	– Herpes 1 and 2 – Influenza – Cytomegalovirus

Table 3.13. *Examples of antiviral compounds characterized in selected species of microalgae and Spirulina (*Spirulina platensis *or* Arthrospira platensis*) (from Dewi* et al. *(2018))*

3.2.3.2. *Antibacterial activities*

Numerous antibacterial activities have been associated with cyanobacteria and microalgae. As with the study of anti-viral activities, antibacterial activities were initially characterized in crude extracts, often of a methanolic nature. Thus, extracts of the green alga *Chlorella vulgaris* showed antibiotic activity against bacteria belonging to the genera *Klebsiella*, *Bacillus* and *Pseudomonas* (Dewi *et al.* 2018). Similarly, methanolic extracts of the microalga *Dunaliella primolecta* revealed a strong antibacterial activity against a strain of *Staphylococcus aureus* resistant to methicillin. The active molecules present in the extracts of these microalgae are mainly phenolic compounds or generally polyunsaturated fatty acids.

Antibacterial activities were also associated with aqueous extracts from cyanobacteria culture. The culture of *Gloeocapsa* sp. cyanobacterium shows activity against bacterial or fungal germs at minimal inhibition concentrations ranging from 3.12 to 12.5 mg/mL culture (see Table 3.14) (Najdenski *et al.* 2013). Aqueous extracts of *Gloeocapsa* show the same type of activity with minimum inhibition concentrations ranging from 1.56 to 6.5 mg/mL of culture depending on the germs considered (see Table 3.14). This activity is mainly carried by the exopolysaccharides since their concentration of inhibition towards the germs tested is clearly lower than that of the raw extracts (see Table 3.14).

Apart from cyanobacteria, aqueous extracts of certain microalgae also have antibiotic activity. Compounds involved in such activities are mainly phycobiliproteins or polysaccharides.

In the red alga *Porphyridium cruentum*, phycobiliproteins inhibit the growth of *S. aureus at a* minimum culture dose of 7 mg/mL. In *P. aerugineum*, this inhibition against the same germ is reduced to 0.29 mg/mL (see Table 3.15) (Najdenski *et al.* 2013). This difference in activity is probably due to the different compositions of phycobiliproteins (C-phycocyanin, B-phycoerythrin) between the two species.

Polysaccharides and more particularly the exopolysaccharides of microalgae, such as cyanobacteria, have also been described as having antibacterial activities. This is notably the case for the species *P. cruentum*, whose sulfated polysaccharides inhibit, at a concentration of 1% (weight/volume of culture), the development of the culture of *Salmonella enteritidis* bacteria, without, however, having an effect on the culture of *S. aureus* (Raposo *et al.* 2014).

Type of extract tested	*Staphylococcus aureus*	*Streptococcus pyogenes*	*Escherichia coli*	*Salmonella typhimurium*	*Candida albicans*
Gloeocapsa sp. (C)	3.12	12.50	12.50	12.50	12.50
Gloeocapsa sp. (AE)	1.56	3.12	6.25	0.00	6.25
Gloeocapsa sp. (EP)	0.125	1.00	0.25	0.25	0.13

Table 3.14. *Examples of antibacterial and antifungal activities present in the cyanobacterium* Gloeocapsa *sp. (C: culture, AE: aqueous extract, EP: exopolysaccharides) expressed as minimum inhibition concentration (mg/mL) (from Najdenski* et al. *(2013))*

Species	***Staphylococcus aureus***	***Streptococcus pyogenes***	***Escherichia coli***	***Salmonella typhimurium***	***Candida albicans***
Arthrospira fusiformis	0.00	0.00	0.00	2.10	2.10
Porphyridium aerugineum	0.29	2.30	0.00	0.00	2.10
Porphyridium cruentum	7.00	7.00	0.00	7.00	7.00

Table 3.15. *Examples of antibacterial and antifungal activities associated with phycobiliproteins from the cyanobacterium of the genus* Arthrospira *and two species of microalgae of the genus* Porphyridium *expressed as minimum inhibition concentration (mg/mL culture) (from Najdenski* et al. *(2013))*

3.2.3.3. *Antitumor activities*

Numerous studies on the preventive role of a diet containing microalgae or cyanobacteria in protecting against cancer have been conducted (Nazih and Bard 2018). The phycobiliproteins, present in cyanobacteria such as Spirulina or red algae belonging to the genus *Porphyridium*, are often cited for their anti-inflammatory, antioxidant and anti-cancerous properties (Fleurence and Levine 2018; Nazih and Bard 2018). C-phycocyanin, a phycobiliprotein produced by Spirulina, has been described as causing apoptosis of malignant cells belonging to the HeLa cell line and colon cancer cells in rats.

In addition to proteins, certain cyanobacterial peptides have also been identified as potential anticancer agents. This is notably the case of a group of peptides or dolostatins present in many cyanobacteria. Apart from cyanobacteria, certain polypeptides synthesized by microalgae are also known for their antitumor activities. In *Chlorella pyrenoidosa* and *Chlorella vulgaris*, a polypeptide, also present in *Spirulina platensis*, has been characterized as having this type of activity.

Cyanobacterial and microalgae polysaccharides, especially exopolysaccharides, also appear to be involved in cancer prevention mechanisms. They are believed to primarily act as immune system activators, helping the host defend itself against the tumor process. This is notably the case of the calcium Spirulina, which is reported to prevent lung metastasis by limiting the adhesion and proliferation of cancer cells (Nazih and Bard 2018).

Besides this, many epidemiological studies highlight the role played by pigments and, more particularly, by carotenoids in the prevention of cancers. Numerous studies carried out *in vitro* show the beneficial effect of this type of pigment in limiting the growth of cancer cells.

Astaxanthin, whose main sources of production are the microalgae *Haematococcus* and *Chlorella*, is one of these pigments with a protective effect. This pigment has been shown to be more effective than other carotenoids (liver, lung).

Other carotenoids, such as fucoxanthin or lutein, are also described as having anticancer activities (Nazih and Bard 2018).

3.2.3.4. *Anti-inflammatory, antioxidant and anti-allergic activities*

Many species of microalgae and some cyanobacteria have been described as having anti-inflammatory, antioxidant and even anti-allergenic activities (Fleurence and Levine 2018). Extracts of the red alga *Porphyridium cruentum* and the green alga *Dunaliella salina* show the inhibitory activity of hyaluronidase, a key enzyme in triggering the inflammatory process (Fujitani *et al.* 2001). Several species of cyanobacteria, including Spirulina, are used in folk medicine for their anti-allergic potential (Kim 2000). In *P. cruentum*, the anti-inflammatory activity is associated with a sulfated polysaccharide with an apparent molecular weight of 500 kDa (Fleurence and Levine 2018). A fragment of this polysaccharide (6.55 kDa) also appears to have strong antioxidant activity. In the alga *Dunaliella bardawil*, the anti-allergic activity and the prevention of asthma induction have been reported to be due to β-carotene. In *Arthrospira maxima* (*Spirulina maxima*), the anti-inflammatory activity and the antioxidant activity are mainly associated with C-phycocyanin (see Figure 3.6). This protein pigment belonging to the phycobiliprotein family is also present in the species *Porphyridium aerugineum* (see section 4.2.1).

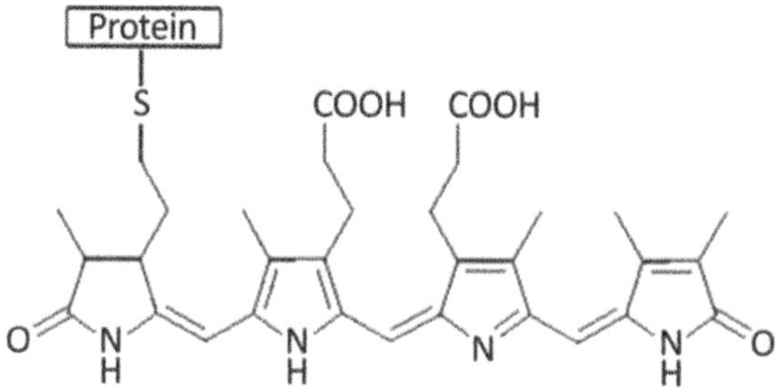

Figure 3.6. *Structure of C-phycocyanin (source: Fleurence and Levine (2018))*

Cyanobacteria and microalgae are endowed with molecules (phycobiliproteins, peptides, exopolysaccharides, pigments) with inert activities from the point of view of human health. These activities have been the subject of a large number of scientific studies. If the claims on human health are difficult to show and delicate to formulate through the concept of "functional food", it remains that the integration of this resource as a dietary supplement in the context of a varied diet is a real nutritional opportunity.

3.2.4. *Food coloring*

The use of microalgae, naturally rich in carotenoid pigments, as colorants in traditional food products, has been the subject of numerous evaluations. The technical feasibility of this approach is often validated, but its economic feasibility remains a barrier to wider development. Two species of the genus *Chlorella*, *Chlorella vulgaris* "green", *Chlorella vulgaris* "orange" (after carotenogenesis), and the species *Haematococcus pluvialis* have been tested for their antioxidant activities and their coloring powers in emulsions (water–oil) (Gouveia *et al.* 2006).

The colors obtained and their intensity are directly related to the biomass introduced and the incorporation rates (0–2% of the volume of the emulsion depending on the species). The most intense orange coloration is observed for the *Haematococcus pluvialis* species. The incorporation of these microalgae at a level of 0.75% in the emulsion generates a much more intense orange coloration than that observed for *Chlorella vulgaris* "orange", even when the latter is incorporated at its maximum rate in the emulsion (1.25%). The integration of "green" *Chlorella vulgaris* colors the emulsion green with an optimal coloring obtained at an integration rate of 2%. The stability of the coloration over time also shows significant interspecific differences. In particular, the coloration generated by the incorporation of *H. pluvialis* appears less stable after six weeks than that produced by the two species of Chlorella. The antioxidant power of the introduced biomasses is also subject to interspecific variations. The three species show a very different antioxidant profile after one and six weeks: the "green" *Chlorella vulgaris* species is the one with the best antioxidant activity for both primary and secondary oxidation mechanisms. This result is linked to the difference in the main pigments present in the three microalgae (astaxanthin, canthaxanthin, lutein).

The species *Chlorella vulgaris* has also been tested to color traditionally formulated butter cookies (Gouveia *et al.* 2007). The addition of algal biomass allowed the accentuation of the green color, which increases with the amount of biomass introduced. This color remains stable during storage (three months at room temperature). The addition of this microalga also increased the firmness of the cakes.

These experiments have shown the benefits of using microalgae and especially their colorants in food formulation. However, they remain limited to technical feasibility, even if some products, such as plankton bread, whose green color of the crumb is probably brought by Chlorella (see Figure 3.3), are subject to seasonal marketing.

3.2.5. *Regulations*

The regulations applied to microalgae and cyanobacteria in human food vary from country to country and also depend on the field of application, i.e. food, ingredient or food supplement.

In France, there are specific regulations on the use of algae (macro- and microalgae) as sea vegetables or food ingredients (Fleurence 2016). This regulation is based on a list of species authorized for human food in the context of this type of application. Initially, this text, established in 1990, applied exclusively to macroalgae and a cyanobacterium, namely Spirulina (*Spirulina* sp.), as well as to its water-extracted pigment C-phycocyanin (Mabeau and Fleurence 1993). Since then, two species of microalgae, *Ondotella aurita* and *Chlorella* sp., have been added to the list. The French regulation around a positive list of authorized algae is, however, more restrictive than the European regulation applied to "Novel Foods and Novel Food Ingredients" of January 27, 1997 (EC Regulation No. 258/97). Indeed, article[1] of this regulation, paragraph 2 d, authorizes the marketing of "foods and food ingredients composed of or isolated from microorganisms, fungi or algae" (EC extract no. 258/97).

Contrary to the French regulation, the European regulation does not establish, through this regulation, a positive list of species of microalgae or cyanobacteria. Moreover, the notion of microorganisms mentioned in the text applies to these last two types of organisms.

Nutritional supplements based on algae are governed by the specific rules for food supplements. As such, they are subject to Directive 2002/46/EC of June 10, 2002 (European Union 2002), which defines food supplements as:

> Foodstuffs the purpose of which is to supplement the normal diet and which are concentrated sources of nutrients or other substances with a nutritional or physiological effect [...] alone

or in combination, marketed in dose form, namely [...] capsules, pastilles, tablets, pills and other similar forms [...] (extract 2002/46/EC).

This definition, therefore, excludes from the application "food supplement", the algal biomass that can be used in the composition of a processed food (e.g. plankton bread) and clearly clarifies the notion of the ingredient of that food supplement.

This European directive has been transcribed into French law, leading to the drafting of a decree specifying the regulatory framework for food supplements (decree no. 2006-352 of March 20, 2006, see (Légifrance 2006)). On the occasion of this decree, the General Directorate for Competition, Consumer Affairs and Fraud Control established a list of microalgae that can be used in the composition of dietary supplements (see Table 3.16) (DGCCRF, SD 4, Secteur Compléments alimentaires, 2019).

Species	**Vernacular name**
Aphanizomenon	Blue-green algae from Klamath Lake
Arthrospira fusiformis, A. indica, A. platensis	Spirulina
Auxenochlorella protothecoides	–
Chlorella sorokiniana, C. vulgaris	Chlorella*
Dunaliella salina	–
Graesiella emersonii	–
Heterochlorella luteoviridis	–
Nannochloropsis oculata	–
Odontella aurita	–
Parachlorella kessleri	–
Scenedesmus vacuolatus	–
Ulkenia sp.	–

Table 3.16. *List of some species of microalgae and cyanobacteria recommended by the DGCCRF for the development of algal-based dietary supplements (based on DGCCRF (2019)). *Author's note: the vernacular name chlorella is not cited in the published list*

It is, however, specified that the "Algae List" is intended to help manufacturers who would like to market food supplements based on algae. It is also specified that this list is not exhaustive and is not legally binding. It is simply recalled that the addition of a new species to this list will require the operator to provide the administration with information on the traditional use of this species in human food.

Concerning regulations outside Europe, there is a wide variety of regulations, or even the absence of regulations, depending on the country. In the United States, since 1994, regulations on dietary supplements have led to two major constraints: precise labeling of the product with its composition and assurance that it does not present a danger to the consumer.

Apart from applications in the field of animal and human nutrition, microalgae and cyanobacteria produce original molecules, such as phycobiliproteins, which are also found in macroalgae, but not in other plants. The production of these molecules, as well as others such as enzymes and polysaccharides, is highly dependent on the environment and therefore on the growing conditions. This metabolic plasticity, associated with the diversity of molecules produced and available for use, is at the origin of the concept of the cell factory frequently attributed to microalgae.

4

Valorized Molecules

Microalgae and cyanobacteria are sources of molecules whose properties can be used in the food, cosmetic or medical fields.

4.1. Polysaccharides

Most of the polysaccharides developing activities that can be valorized are polysaccharides produced by the cell membrane and are in contact with the external environment. Under certain conditions, they can also be released into the environment. These saccharide polymers are called exopolysaccharides (EPS). In the natural environment, these exopolysaccharides protect the organism from desiccation, bacterial infections and attack by predatory protozoa. In cyanobacteria, exopolysaccharides also contribute to the adhesion of these organisms on a surface and play an important role in the production of biofilm. These physical and biological properties are at the origin of the interest in polysaccharide enhancement.

In cyanobacteria, exopolysaccharides are characterized by a wide biochemical variety (two to ten different types) and by the nature of their constituent monomers. These monosaccharides or bones are, in general, neutral or acidic sugars. These acid sugars, such as uronic acids, and more particularly glucuronic acid (see Figure 4.1), will give the polymer an anionic character. Other charged groups, such as pyruvate or sulfate groups, will generate an anionic charge to the polysaccharides. Polypeptide or even acetyl substituents can also sometimes be observed, making the structure of some exopolysaccharides relatively complex (Philippis and Vincenzini 1998). These exopolysaccharides have a very high affinity for water and have viscosity properties that make them potential processing aids (additives) for industry.

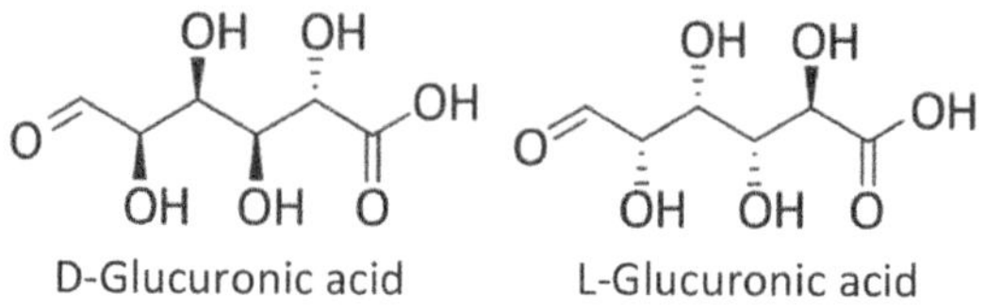

D-Glucuronic acid L-Glucuronic acid

Figure 4.1. *Structure of glucuronic acid (D and L isomers) (source: Grovel 2020)*

In the common *Nostoc* cyanobacterium, exopolysaccharides consist of monosaccharides of neutral sugars: glucose, xylose and galactose (Han *et al.* 2013). In the natural environment, glucose represents 50% of the monomeric composition of the polysaccharides of this cyanobacterium (see Table 4.1) (Huang *et al.* 1998). In a culture, in an artificial medium containing nitrogen or not, the glucose content represents, respectively, 7.1 and 14% of the total osidic composition. EPS produced in a culture medium without nitrogen are characterized by high levels of glucose (48.9%), arabinose (25.7%), and to a lesser extent, xylose (12.0%). In a nitrogen-containing culture medium, the exopolysaccharides produced are characterized by the majority presence of xylose (40.1%) and galactose (18.8%) (see Table 4.1). Other experiments, carried out in natural or artificial environments, have confirmed the role played by the environment on the composition of the monomers constituting the exopolysaccharides of this cyanobacterium. Thus, Brüll *et al.* (2000) have showed that the polysaccharides produced by Nostoc, grown in a medium not limited in nitrogen, are distinguished as follows from those generated in natural or low-nitrogen environments. They are characterized in particular by the presence of high levels of 2-O-methyl glucose and glucuronic acid. In nitrogen-poor environments, the exopolysaccharides produced have arabinose contents much higher than those synthesized in non-nitrogen-limited environments. This metabolic plasticity, as well as the interspecific variability, leads to the production of molecules with different physicochemical or biological properties.

Nostoc	Arabinose (%)	Xylose (%)	Galactose (%)	Glucose (%)
Natural environment	–	27.0	21.8	50.2
N-rich medium culture	6.3	43.5	21.2	7.1
N-poor culture	1.8	35.1	20.2	14.0
EPS N-rich medium	9.5	40.1	18.8	12.5
EPS N-poor environment	25.7	12.0	11.7	48.9

Table 4.1. *Relative composition of the main oses of the constituent monosaccharides of polysaccharides and exopolysaccharides of the common* Nostoc *cyanobacterium as a function of the medium (from Huang* et al. *(1998))*

Among the physicochemical properties associated with the exopolysaccharides of cyanobacteria is mainly viscosity. This is a function of the biochemical nature of the exopolysaccharides, which varies, for a given species, from culture conditions and also from species.

The species *Cyanospira capsulata* is known to have a pseudoplastic behavior during its culture, especially at the beginning of growth. This is due to the exopolysaccharides and their high viscosity. At the end of cultivation, viscosity decreases as shear forces increase (Philippis and Vincenzini 1998). The polysaccharides produced by this cyanobacterium, with a content of 0.1% w/v, show a rheological behavior slightly superior to that of xanthan gum which is a gelling agent (E415) frequently used in the food industry.

In addition to the applications of cyanobacterial exopolysaccharides as potential gelling agents, there are also opportunities for valorization in the medical field. As previously discussed (see section 3.2.3), cyanobacterial polysaccharides are associated with many therapeutic activities. For example, the acid fractions of common *Nostoc* polysaccharides exhibit complement fixation inhibitory activity of up to 90% at a concentration of 500 µg/mL (Brüll *et al.* 2000). This suggests a significant anti-inflammatory activity associated with this type of molecule. In other species such as *Arthrospira platensis*, exopolysaccharides have shown antiviral activity against herpes virus and acquired immunodeficiency virus (HIV) (Philippis and Vincenzini 1998). The exopolysaccharide calcium spirulan, produced by *Arthrospira platensis* (spirulina), has been described as inhibiting the replication of many viruses, such as those of herpes, measles and mumps (Hayashi *et al.* 1996).

This sulfated polysaccharide also appears to be active in inhibiting the penetration of the virus into the host cell.

In microalgae, EPS are also described as having rheological and biological activities that can be used for industrial applications. The red microalga *Porphyridium cruentum* is known to produce non-dialyzable, i.e. high molecular weight, sulfated expolysaccharides (Medcalf *et al.* 1975). Typical polysaccharides are described as containing 7.4% sulfate groups, glucuronic acid (9%), xylose and galactose with an almost equivalent molar ratio (1/0.9). A small proportion of glucose and traces of mannose and rhamnose have also been reported by Medcalf *et al.* (1975). The viscometric properties of *P. cruentum* polymers have also been extensively studied. They show an aptitude to form very viscous solutions for a very low level of presence in solution (1%). This viscosity increases exponentially between 1 and 2%, just like guar gum. Furthermore, the viscosity is pH-dependent. Indeed, it appears maximum for pH between 2 and 6 and becomes minimum for basic pH (11–12). The viscosity is also different according to the nature and concentration of the salts present in the solution. It is higher at a salt concentration of 5% and in the presence of NaCl. However, at this concentration, the viscosity is lowered by 18.50% in the presence of $CaCl_2$. Temperature also plays a role in the viscosity properties of *P. cruentum* polysaccharides. Between 15 and 70°C, the viscosity decreases linearly and then rises to a maximum when the temperature of the solution is raised to 80°C.

Recent work, based on the use of membrane filtration, has been used to refine the apparent molecular weight of *P. cruentum* polysaccharides (Marcati *et al.* 2014). Nearly 22% of the polysaccharides are retained by a membrane with a cut-off zone of 300,000 daltons. Nearly 7% of the polysaccharides are retained by a membrane whose cut-off zone is established at 10,000 daltons. 70% of the remaining polysaccharides have molecular weights between these two limits. The exopolysaccharides retained by the membrane with a cut-off of 300,000 daltons therefore represent very high molecular weight heteropolymers. The latter is between 2 and 6×10^6 daltons, according to some authors (Sun *et al.* 2009). For others, the average is around 2.3×10^6 daltons, with a mass distribution between 0.2 and 4×10^6 daltons (Geresh *et al.* 2002).

P. cruetum exopolysaccharides are also tested for their therapeutic activities. In particular, they have antiviral and antibacterial activities (see sections 3.2.3.1 and 3.2.3.2). The degree of sulfation of the polysaccharides seems to play a major role in the efficacy of their antiviral activity. Other factors such as the molecular weight of the polymer and the stereochemical position of the sulfated groups are also advanced (Raposo *et al.* 2014). The competitive binding of exopolysaccharides to the cell sites recognized by the viral envelope glycoproteins could limit the penetration of the virus into the host cell, as described for calcium spirulan. The polyanionic load of *P. cruentum* polysaccharides could also interact with the positive loads of the amino acids present on the viral envelope and, more particularly, on those constitutive of the virus attachment glycoproteins to the cell (Raposo *et al.* 2014).

The antibacterial activity of *P. cruentum* exopolysaccharides appears to be related to the physicochemical structure of these polymers as well as to the nature of the bacteria wall. Their antibacterial properties seem to depend on their ability to inhibit the formation of a bacterial biofilm and thus limit the bacteria's ability to bind. This mechanism would explain the difference in the activity of exopolysaccharides on Gram (–) or Gram (+) bacteria, which differ from each other by the composition of their wall.

In addition to their antiviral and antibacterial properties, antioxidant activities have also been associated with the exopolysaccharides of *P. cruentum* (see section 3.2.3.4). This activity differs according to the size of the polysaccharides. It is very important for 6.55 kDa fragments and lower for 60 and 256 kDa fragments (Sun *et al.* 2009).

Other microalgae also produce exopolysaccharides of interest in terms of recovery. The EPS of the algae *Chlorella pyrenoidosa*, *Chlorococcum* sp. and *Scenedesmus* sp. have recently been characterized (Zhang *et al.* 2019). These polysaccharides have very different molecular weights depending on the species (see Table 4.2).

	Chlorella pyrenoidosa	***Chlorococcum* sp.**	***Scenedesmus* sp.**
Molecular mass of exopolysaccharides (daltons)	194,000	32,400	7,390

Table 4.2. *Apparent molecular mass of major exopolysaccharides of the microalgae* Chlorella pyrenoidosa, Chlorococcum *sp. and* Scenedesmus *sp. (from Zhang* et al. *(2019))*

The osidic composition of these exopolysaccharides also shows significant differences (see Table 4.3). The exopolysaccharides of *Chlorococcum* sp. and *Scenedesmus* sp. are distinguished from those of *Chlorella* by the majority presence of glucose (+30%) and glucosamine (+20%). Conversely, the exopolysaccharides of Chlorella are characterized by the majority presence of galactose and arabinose.

The biological activities carried by these polysaccharides have also been evaluated, most of them *in vitro*.

The antioxidant properties of the exopolysaccharides of these three species and, more particularly, their capacity to neutralize free radicals, agents causing cellular and molecular damage (proteins, DNA), were tested. *Scenedesmus* exopolysaccharides at a concentration of 1 mg/mL proved to be the most effective, neutralizing nearly 75% of free radicals (Zhang *et al.* 2019) (see Table 4.4). Antitumor activities against two colon cancer cell lines (A HCT116, B HCT8) were also evaluated. Cell viability of the A HCT116 cell line is reduced to nearly 80% after 24 hours at an exopolysaccharide exposure dose of 0.6 mg/mL, regardless of the algal species tested (Zhang *et al.* 2019). Under the same conditions, the exopolysaccharides of *Chlorella*, *Scenedesmus* and *Chlorococcum* limit the viability of B lineage HCT8 cells to 64, 61 and 77%, respectively. These results suggest that the exopolysaccharides of these three microalgae species are interesting candidates for the development of antitumor agents.

	Chlorella pyrenoidosa	***Chlorococcum* sp.**	***Scenedesmus* sp.**
Galactose (%)	37.1	15.6	8.4
Glucose (%)	8.4	33.0	39.0
Glucosamine (%)	13.5	21.9	28.8
Arabinose (%)	19.0	2.0	0.5
Mannose (%)	7.0	17.6	17.3

Table 4.3. *Partial osidic composition of exopolysaccharides of the microalgae* Chlorella pyrenoidosa, Chlorococcum *sp. and* Scenedesmus *sp. (from Zhang* et al. *(2019))*

	0 mg/mL	0.2 mg/mL	0.4 mg/mL	0.6 mg/mL	0.8 mg/mL	1.0 mg/mL
% free radical neutralization	0	27	48	65	70	75

Table 4.4. *Percentage of free radical neutralization by* Scenedesmus *sp. exopolysaccharides as a function of concentration (from Zhang* et al. *(2019))*

Some cyanobacteria and microalgae produce charged exopolysaccharides, most often sulfated, whose composition varies according to the species, as well as to the environmental conditions, i.e. culture conditions. This adaptation in the production of molecules of interest is at the origin of the "cell factory" concept applied to microalgae.

4.2. Proteins and enzymes

Cyanobacteria and microalgae, like all organisms, produce proteins. A distinction must be made between proteins that play a structural role of those that act as biochemical catalysts, such as enzymes. Some proteins are very original because their presence is limited to cyanobacteria and algae (micro and macroalgae). These are more particularly the phycobiliproteins that act as photon-collecting antennae and participate with the chlorophyll in the photosynthetic process. At the enzymatic level, the enzymes involved in free radical scavenging mechanisms, such as superoxide dismutase (SOD), play a fundamental role in the adaptation of these microorganisms to the oxidizing conditions of the environment. The synthesis of phycobiliproteins and SOD and their activities are highly dependent on the environment. This reminds us that the concept of cell factory does not only apply to the production of polysaccharides.

4.2.1. *Phycobiliproteins*

Phycobiliproteins are macromolecules consisting of a protein or apoprotein part and a non-protein part consisting of an open tetrapyrrole nucleus (see Figure 3.6). This tetrapyrrole nucleus acts as a photon collecting antenna (see Figure 4.2) and transfers 90% of the collected light energy as fluorescence to the chlorophyll-containing photosystem. Within the cell, phycobiliproteins are associated with each other in a supramolecular complex known as phycobilisome (see Figure 4.3). There are three main categories of phycobiliproteins, namely, allophycocyanin, phycocyanin and phycoerythrin (see Figure 4.4). Each of these protein pigments is characterized by absorption wavelengths and fluorescence emission wavelengths (see Table 4.5). These spectral properties, and more particularly their ability to emit fluorescence, are at the origin of the development of phycobiliproteins. Their production by the cell is highly dependent on light intensity and photoperiod, which makes these macromolecules easily inducible compounds under culture conditions. Thus, in spirulina, for a biomass of 0.24 g/L, maximum C-phycocyanin production is obtained when the culture is subjected to a light intensity of 300 mmol/m^2/s (Xie *et al.* 2015).

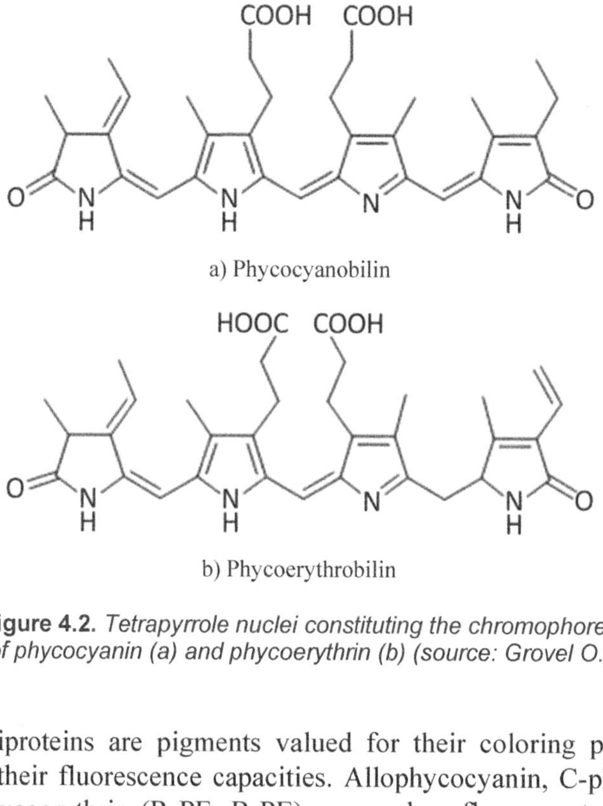

a) Phycocyanobilin

b) Phycoerythrobilin

Figure 4.2. *Tetrapyrrole nuclei constituting the chromophores of phycocyanin (a) and phycoerythrin (b) (source: Grovel O.)*

Phycobiliproteins are pigments valued for their coloring properties as well as for their fluorescence capacities. Allophycocyanin, C-phycocyanin, R and B-phycoerythrin (R-PE, B-PE) are used as fluorescent reagents for cell typing of B or T lymphocytes and in antibody labeling in immunofluorescence reactions. R-phycoerythrin is widely used in many techniques based on fluorescence detection. This includes the calibration of flow cytometers, ELISA or immunofluorescence techniques, as well as other analytical applications where high sensitivity without photostability is required. The marketing price of this molecule is 152 euros per mg, making R-Phycoerythrin a molecule with very high added value (Sigma-Aldrich 2020).

Figure 4.3. *Phycobilisome (source: Pouchus Y.-F.). For a color version of this figure, see www.iste.co.uk/fleurence/microalgae*

	Color	λmax absorption (nm)	Secondary absorption (nm)	Fluorescence emission (nm)	Algae
Allophycocyanin	Blue	618/671	–	675	Red
C-Phycocyanin	Blue	520	–	640	Cyanobacteria
R-Phycocyanin	Blue	555/617	–	636	Red
R-Phycoerythrin	Pink	540/565	–	575	Red
B-Phycoerythrin	Pink	545/565	498	576	Red

Table 4.5. *Spectral characteristics of major phycobiliproteins of red microalgae and cyanobacteria (from Morançais* et al. *(2018))*

Figure 4.4. *Phycoerythrin (photo credit © Fleurence J., 2005). For a color version of this figure, see www.iste.co.uk/fleurence/microalgae*

Apart from their applications as fluorescent probes, phycobiliproteins are also valued as food colorants. This is notably the case of C-phycocyanin, which, extracted from *Arthrospira platensis* (spirulina), is marketed in China and Japan under the name Lina Blue (Dumay *et al.* 2014). Lina Blue is frequently used to color ice cream, sorbets, candies, beverages and dairy products.

A red microalgae *Porphyridium aerugineum* has the characteristic to produce a blue dye closer to C-phycocyanin than to the R-phycocyanin, which is the pigment that accompanies phycoerythrin in red algae. This pigment is very stable in culinary preparations such as sweet flowers to decorate cakes and shows a color stability of several years when stored under normal conditions (Dufossé *et al.* 2005). It is also suitable for more complex formulations such as gelatin and ice cream. Despite this, it is not commercially produced, as its regulatory status as a food additive is not clearly regulated.

The use of phycoerythrins (R or B) as red food colorants is technically feasible. However, their use for this type of application is strongly constrained by European and French regulations, which limit their development. Despite this, Taiwan markets an extract enriched in R-phycoerythrin, which is distributed as a food colorant for the Asian market (personal communication, Fleurence J.).

Phycobiliproteins have also been described as developing antitumor properties (see section 3.2.3.3). Apart from phycocyanins, phycoerythrins have also been reported to have this type of activity. This is particularly the case for B-phycoerythrin produced by the microalga *Porphyridium cruentum* (Minkova *et al.* 2011). This pigment appears to be particularly active against Graffi's myeloid tumor, which is classified as a rare form of leukemia. At a dose of 50 μg and 100 μg/mL, B-phycoerythrin from this alga inhibits, respectively, 50 and 63% of tumor cell growth obtained from the hamster model (Minkova *et al.* 2011). This pigment notably promotes the apoptosis of malignant cells. However, its action is not limited to the inhibition of tumor growth. It is also involved in the stimulation of normal bone marrow stem cells. Indeed, it has a stimulating activity leading to the proliferation of these cells of the order of 154% for a dose of 100 μg/mL (Minkova *et al.* 2011). This dual activity makes B-phycoerythrin a molecule of interest for the development of therapeutic tools for this orphan disease.

4.2.2. *Enzymes*

Cyanobacteria and microalgae have developed enzymatic systems that enable them to resist the oxidative stresses of the environment. Among these enzymes are mainly superoxide dismutase (SOD), catalase and peroxidase.

In particular, oxidative stress produces superoxide ions that act as extremely reactive free radicals that can react with biological membranes and certain molecules of vital interest to organisms, such as DNA.

The SOD will proceed with the dismutation of the superoxide anions into oxygen and hydrogen peroxide molecules (hydrogen peroxide) via the following enzymatic reaction:

$$2\ O_2^{-} + 2\ H^{+} \rightarrow O_2 + H_2O_2$$

The catalase will then neutralize the hydrogen peroxide, a toxic species for living organisms, through the following reaction:

$$2\ H_2O_2 \rightarrow 2\ H_2O + O_2$$

The neutralization reaction of hydrogen peroxide can also take place via peroxidases through the following reaction:

$$RH_2 + H_2O_2 \rightarrow R + 2\ H_2O$$

Superoxide dismutases are, therefore, detoxification enzymes with high antioxidant power. In *Arthrospira platensis*, SOD activity is inducible over time during exposure to oxidative stress. This induction occurs at the level of expression of the gene coding for SOD with a maximal effect after 10 days of exposure (Sannasimuthu *et al.* 2018). The antioxidant activity of SOD is mainly carried by a peptide, L112L, which has a catalytic site. This peptide at a concentration of 12.5 μM considerably reduces the activity of superoxide anions and shows no cytotoxicity towards human cells such as leukocytes. The L112L peptide, therefore, appears to be a potentially interesting molecule for the development of drugs with antioxidant activities.

Although all SODs catalyze the same type of reaction, there is a wide variety of enzymes in microalgae, which are distinguished by their molecular weight, stability and physicochemical characteristics (see Table 4.6).

Most of them are macromolecules with molecular weights ranging from 30,000 to 80,000 daltons with manganese as a cofactor and an isoelectric point of 4.2 (Gudin and Trezzy 1994). In the red microalgae *Porphyridium cruentum*, SOD is an enzyme with an apparent molecular weight of 40,000 daltons. It consists of two subunits of equivalent molecular mass (Misra and Fridovich 1977; Zeinali *et al.* 2015). The SOD of the diatom *Thalassiosira weissflogii* is characterized by relatively interesting physicochemical characteristics. It is thermostable, and its activity can withstand pH variations from 4 to 12. Moreover, this enzyme is also resistant to the presence of proteases (Zeinali *et al.* 2015). These rather remarkable properties make SOD a potential antioxidant active ingredient for industry, which is always looking for enzymes that can withstand the constraints of food or cosmetic formulation.

Induction of the activity or synthesis of SOD and catalases under the influence of oxidative or light stress also has the advantage of stimulating the production of these enzymes in a controlled environment. This has been particularly demonstrated for the green microalgae *Chlorella vulgaris*. In this species, irrigating the crop with doses of ultraviolet B rays increases the production of superoxide ions by 284% and lipid peroxidation by 145% (Malanga and Puntarulo 1995). The activities of SOD and of catalase are, in the latter case, increased by 40 and 500%, respectively.

Species	Group	Temperature (°C)	pH
Cyanidium caldarium	Red algae	40–50	1–5
Chlorella vulgaris	Green algae	40–50	7
Chlorella saccharophila	Green algae	40–60	7
Dunaliella salina	Green algae	40–50	7
Synechococcus lividus	Cyanobacteria	50–80	7
Synechococcus elongatus	Cyanobacteria	50–70	7

Table 4.6. *SOD stability as a function of temperature and pH for selected species of microalgae and cyanobacteria (from Gudin and Trezzy (1994))*

In the cosmetic field, SOD and catalase are mainly valorized indirectly through hydro glycolic extracts obtained from microalgae. In 1994, a patent was filed for the use of an extract composed of green coffee, spirulina, and the microalgae *Chlorella* sp. and *Scenedesmus* sp. (Brin and Goutelard 1994).

The antioxidant activity of the extract is, however, lower than that measured for a purified SOD. Nevertheless, the use of purified SOD or catalase produced by algal biomass, even under favorable conditions, is essentially opposed to economic or regulatory feasibility. This observation is particularly relevant for sectors of activity with low- or medium-added value, such as food and cosmetics.

4.3. Non-protein pigments

In addition to phycobiliproteins, cyanobacteria and microalgae have non-protein pigments whose synthesis can be modulated by the surrounding environment and, therefore, by the culture conditions. Among the latter, a distinction must be made between chlorophyll pigments and carotenoid pigments. In cyanobacteria and microalgae, the routes of valorization mainly concern carotenoids. These pigments are mainly used as food colorants (see sections 3.1.2.5 and 3.2.4). In addition to their coloring properties, carotenoids have activities of interest for human health, such as antitumor activities (see section 3.2.3.3).

In the microalgae *Chlorella ellipsoidea* and *Chlorella vulgaris*, antiproliferative activity of carotenoids on colon cancer cell line HCT 116 have been detected (Cha *et al.* 2008). This activity appears to be associated with violaxanthin for *C. ellipsoidea* and lutein for *C. vulgaris*. The concentration for 50% inhibition (CI 50) of malignant cell growth is 40.7 μg/mL and 40.3 μg/mL, respectively, for each of the two pigments. However, *C. ellipsoidea* extract, particularly rich in violaxanthin, shows an apoptosis-inducing activity of cancer cells 2.5 times higher than that noted for *C. vulgaris* extract containing mainly lutein. The first pigment mentioned is related to xanthophylls and the second to β-carotene. These kinds of results suggest that xanthophyll-type pigments would have a higher antitumor activity than carotene-type pigments. Fucoxanthin, a xanthophyll pigment of the carotenoid family, has also been described for its antitumor activity on bronchopulmonary and epithelial cancer cell lines. This pigment was obtained from extracts of the diatom *Odontella aurita* and the haptophyte *Isochrysis galbana* (Moreau *et al.* 2006). The microalga *Dunaliella salina*, cultivated for its ability to accumulate carotenoids, has also been the subject of studies on the biological properties of these pigments in the field of health. In particular, the *in vivo* antioxidant activities of *D. salina* pigments, especially carotenoids, have been well described

(Murthy *et al.* 2005). In the albino rat model subjected to oxidative stress via exposure to tetrachloromethane (CCl4), the administration of an algal extract at doses of 125 µg/kg and 250 µg/kg body weight shows a real protective effect on the animals with regard to the toxicity induced by CCl4. In particular, the administration of carotenoids would allow the restoration of SOD, catalase and peroxidase activities, thus limiting the effects of oxidative stress linked to the addition of a toxic solvent. However, the antioxidant activity of non-protein pigments is not the only interest of carotenoids.

The diatom *Haslea ostrearia*, well known in the phenomenon of oyster greening, produces an original pigment of polyphenolic nature, whose structure is not yet elucidated and whose biological role is far from being totally defined. This water-soluble blue-green pigment is concentrated at the two apical poles of the microalgae (see Figure 1.6). This pigment called marennine exists in an intracellular form or "internal marennine" (Mn I) and in an extracellular form or "external marennine" (Mn E). Both forms of marennine are very effective in trapping superoxide anions (Pouvreau *et al.* 2008). At a concentration of 10 µmol/L, Mn I and Mn E neutralize 80% and 90% of the superoxide ions present in the medium, respectively. A 100% trapping rate is observed for both forms of marennine when present at 100 µmol/L concentration. Marennine, whatever its form, is equally effective in neutralizing free radicals generated in the medium. Thus, at a concentration of 50 µmol/L, Mn I and Mn E trap 70% and 90% of free radicals, respectively. Mn E, excreted by the algae, proves to have significantly higher antioxidant properties than those reported for Mn I. This last characteristic is interesting in terms of recovery since Mn E, unlike Mn I, does not require an extraction step. Indeed, the extraction operation is carried out by the algae itself through the secretion process. Marennine, a pigment whose color varies from blue to green depending on the pH and whose antioxidant activities are very high, represents an opportunity for certain value as a dye and antioxidant. Its authorization as a food additive may, however, be difficult since the structure of the molecule is not yet elucidated (personal communication, Fleurence J.).

4.4. Fat, sterols and fatty acids

As in many plants, the lipids of microalgae are present in polar or apolar form (neutral lipids). Polar lipids or polar glycerolipids consist of phospholipids and glycolipids (see Figure 4.5). These lipids constitute between 40 and 92% of the total lipid fraction (Mimouni *et al.* 2018). Conversely, non-polar lipids, apolar glycerolipids, represent between 5 and 51% of the lipid fraction. However, the total lipid content varies according to species, growth stages and environmental conditions. As for the compounds previously described, lipids are macromolecules whose production is also induced by culture conditions.

Neutral lipids include apolar glycerolipids as well as sterols and free fatty acids. Diatoms and chlorophyceae are the algae that contain the highest levels of neutral lipids (Mimouni *et al.* 2018). Total fatty acid levels also vary greatly between microalgae species. Some chlorophyceae or dinophyceae have particularly high levels of total fatty acids, up to 48 and 188 picograms/cell, respectively (see Table 4.7) (Viso and Marty 1993). Fatty acids are classically classified as saturated fatty acids (absence of double bonds), monounsaturated fatty acids (presence of a single bond) and polyunsaturated fatty acids (presence of two or more double bonds). Polyunsaturated fatty acids are conventionally classified as n-3 or n-6 acids depending on the position of the carbon where the first double bond is located (see Figure 4.6). The distribution of saturated fatty acids/mono- or poly-unsaturated acids varies greatly depending on the species (see Table 4.8).

The production of lipids such as triacylglycerols (see Figure 4.5) or polyunsaturated fatty acids can depend on many factors, such as light, salinity, temperature or the availability of nitrogen in the surrounding environment. When growing the microalga *Nannochloropsis* sp. the increase in light intensity, salinity and nitrogen supply increases the production of triacylglycerols (Pal *et al.* 2011). Under certain stressful conditions, including high salinity, a reduction in the production of eicosapentaenoic acid is observed. On the contrary, moderate salinity (13 g/L NaCl) results in a total fatty acid fraction of nearly 28% eicosapentaenoic acid, which represents only 22% of the total fatty acid fraction when the alga is cultured in a nitrogen-depleted and hypersaline medium (40 g/L). In the diatom *Ondotella aurita*, temperature is a factor strongly influencing the production of polyunsaturated fatty acids such as arachidonic acids and eicosapentaenoic acids (see Table 4.9). Thus, the eicosapentaenoic acid content is maximum at 8°C (38.91% of the total fatty acid fraction) and minimum at 24°C (18.31% of the total fatty acid fraction) (Pasquet *et al.* 2014).

In the microalgae *Porphyridium cruentum*, under salinity conditions of 28 g/L NaCl and a culture pH of 8, the optimal temperature for obtaining polyunsaturated fatty acids is 18°C. Under these conditions, fatty acids represent 43.7% of the total fatty acid fraction, and eicosapentaenoic acid is dominantly present, constituting nearly 25.5% of the polyunsaturated fatty acids (Durmaz *et al.* 2007).

Light is also an important factor influencing the production of arachidonic acid. In the freshwater microalgae *Parietochloris incisa*, a light intensity of 400 μmol photons/m^2/s generates arachidonic acid production of about 85 mg/L (Solovchenko *et al.* 2008). This production is 27 mg/L for an irradiance of 35 μmol photons/m^2/s.

In terms of valorization in nutrition and human health, polyunsaturated fatty acids are those of major interest, especially eicosapentaenoic, arachidonic and docosahexaenoic acids (DHA). The latter are often described for their anti-inflammatory, antihypertensive, antioxidant, immunoregulatory, antithrombotic and cardiovascular protective activities (Mimouni *et al.* 2018).

Polar lipids and, more particularly, glycolipids have often been associated with activities of interest to human health. In the freshwater alga, *Chlorella vulgaris*, a novel monogalactosyldiacylglycerol (see Figure 4.5) has been identified as an inhibitor of tumor cell promoters (Morimoto *et al.* 1995). Other glycolipids are also known for their angiogenesis inhibition activities and their antiviral or anti-inflammatory properties (Mimouni *et al.* 2018). Antiviral, antitumor and antibacterial properties are also associated with the phosphor lipids of microalgae.

The lipids and fatty acids of microalgae are a source of valuable molecules in the fields of nutrition and human health. Their production varies according to species as well as according to environmental parameters such as light, temperature or salinity. Their synthesis is therefore easily programmable, at least in theory. This lipid plasticity is at the origin of the production of fatty acids with a long carbon chain and has initiated another way of enhancing the value of microalgae. This new pathway, which aims to produce biofuels based on microalgae, is currently undergoing rapid expansion.

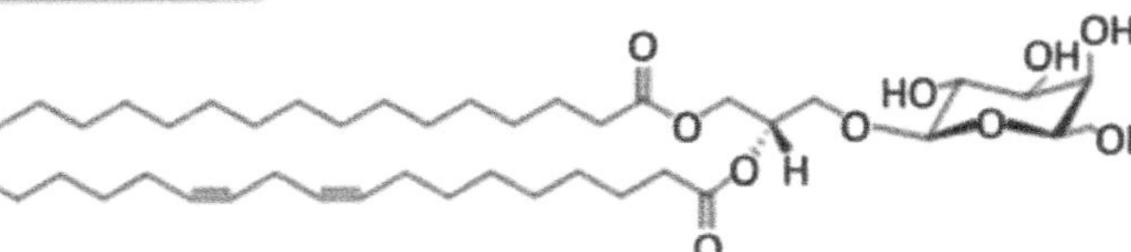

Figure 4.5. *Examples of lipids belonging to the different classes (neutral lipids, glycolipids and phospholipids) (source: Grovel O.)*

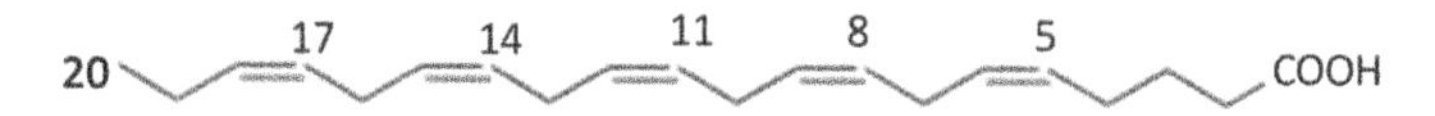

Figure 4.6. *Example of n-3 polyunsaturated fatty acid (eicosapentaenoic acid) (source: Grovel O.)*

Species	Family	Total fatty acids (pg/cell)
Nitzschia closterium	Diatom	12.9
Prorocentrum micans	Dinophyceae	109.3
Scrippsiella trochoidea	Dinophyceae	187.7
Dunaliella primolecta	Chlorophyceae	48.1
Porphyridium cruentum	Rhodophyceae	3.2

Table 4.7. *Examples of total fatty acid contents of some microalgae species (expressed in picograms per cell) (from Viso and Marty (1993))*

Species	16:0	16:1	18:1	18:2	18:3 n-6	18:4 n-3	20:4 n-6	20:5 n-3
Thalassiosira pseudonana	10	29	–	1	–	–	14	15
Parietochloris incisa	10	2	16	17	1	–	43	1
Amphidinium carteri	12	1	2	1	3	19	20	–
Ochromonas danica	4	–	7	26	12	7	8	–
Porphyridium cruentum	34	1	2	12	–	–	40	7

Table 4.8. *Major fatty acids in selected microalgae relatively rich in arachidonic acid (C20:4) (expressed as % of total fatty acids) (based on Cohen and Khozin-Goldberg (2010))*

4.5. The special case of biofuel

4.5.1. *Biofuel production processes*

Many terrestrial plants such as soybean, sunflower or rapeseed are already used for the production of biogas or biofuels. The same type of concept applies to algal biomass, which can be transformed for the production of biofuel in liquid (oil and alcohol), solid (coal) or gaseous form. The main types of biofuel are methane, hydrogen, ethanol, oils and diesel. Biomethane is the result of the conversion of organic matter, bioethanol from the alcoholic fermentation of sugars and biodiesel from the conversion of the lipid fraction, i.e. algal oil.

The transformation of algal biomass into fuel is based on two totally different processes (see Figure 4.7). The first conversion process is thermochemical, and the second is biochemical (Brennan and Owende 2010). The thermochemical process is based on the decomposition of organic matter at high temperatures (300–1,000°C) to produce biofuel. This process can be carried out in four different ways:

– gasification;

– thermochemical liquefaction;

– pyrolysis;

– direct combustion.

In the gasification process, the biomass is subjected to combustion via a gas raised to a very high temperature (800–1,000°C). The combustion of the biomass will generate a gas mixture consisting mainly of hydrogen, carbon monoxide and dioxide, and methane or biomethane. This synthetic gas is called syngas or town gas in popular parlance. However, it is poorer than natural gas from an energy point of view.

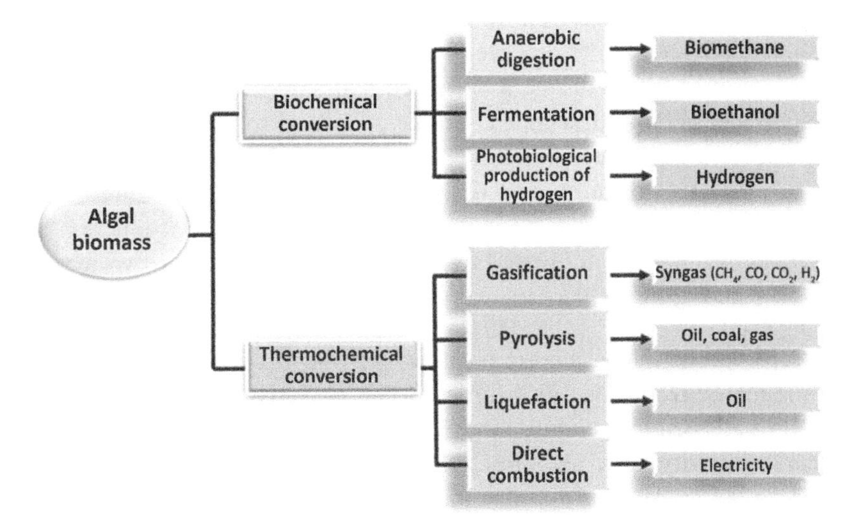

Figure 4.7. *Main processes and procedures used for the conversion of algal biomass to biofuel (source: Pouchus Y.-F., based on Brennan and Owende (2010)). For a color version of this figure, see www.iste.co.uk/fleurence/microalgae*

This process has been tested on spirulina. The conversion rate is 0.64 g of methanol produced for 1 g of gasified biomass (Brennan and Owende 2010).

Thermochemical liquefaction is a process that operates at lower temperatures than the previous process, between 300 and 350°C. However, it requires the use of high pressure (5–20 MP) and the presence of hydrogen as a chemical catalyst. In this process, the algal biomass is directly transformed, not into gas but into liquid fuel.

In the green alga *Botryococcus braunii*, this process allows the conversion of 64% of the biomass, expressed in terms of dry matter, into fuel oil. The energy obtained at the end of this process is approximately 46 MJ/kg. Another study carried out on the *Dunaliella tertiolecta* species shows that 42% of the algal biomass, expressed in relation to dry matter, is converted into oil with this process (Brennan and Owende 2010). The energy recovered is then 35 MJ/kg.

Pyrolysis is an operation that allows the transformation of algal biomass into oil, syngas and coal in the absence of air and at temperatures between 350 and 700°C. Three types of pyrolysis are applicable to algal biomass depending on the end product that is mainly sought after (oil, coal or gas) (see Figure 4.8). These are flash pyrolysis, fast pyrolysis and slow pyrolysis. Flash pyrolysis is characterized by the application of a temperature of 500°C and a hot steam stream for one second. Fast pyrolysis is also performed at a temperature of 500°C with the addition of a stream of moderate heat steam over a period of 10–20 seconds. Finally, slow pyrolysis is performed at 400°C with a long heating time. The flash pyrolysis and the fast pyrolysis are to be preferred for the production of oils from biomass (Figure 4.7). On the contrary, flash pyrolysis seems to be unsuitable for gas production, as it generates only 13% gas compared to 30 and 35% for the other two pyrolysis processes (see Figure 4.8).

Direct combustion is a process in which the biomass is incinerated at temperatures above 800°C in a furnace. The objective of this process is to transform the chemical energy stored in the biomass into hot gases and to produce electricity. This is only feasible if the moisture content of the biomass is below 50% of the dry matter. The main disadvantage of direct combustion is that it requires prior treatment of the biomass, such as drying, chopping and grinding, which leads to additional energy costs for the process.

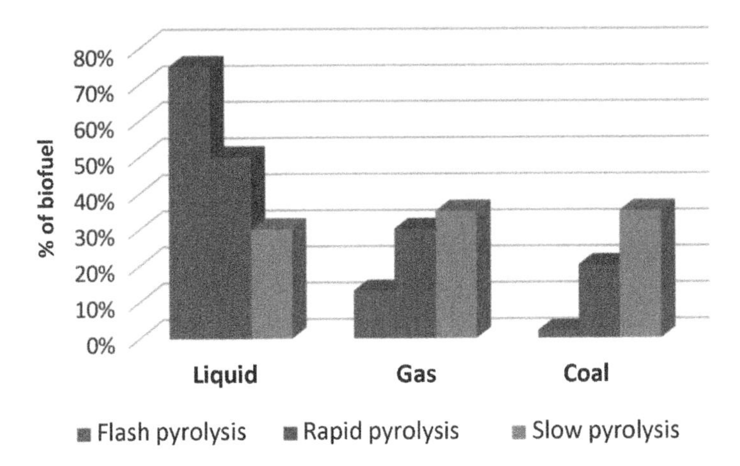

Figure 4.8. *Distribution by product type (liquid, gas and coal) of the different pyrolysis processes applied to algal biomass (based on Brennan and Owende (2010)). For a color version of this figure, see www.iste.co.uk/fleurence/microalgae*

The process of biochemical conversion of algal biomass is based on three distinct processes: anaerobic digestion, alcoholic fermentation and photobiological hydrogen production.

The anaerobic digestion of algal biomass generates a gas mainly composed of methane (CH_4) or biomethane and carbon dioxide (CO_2). This process has the advantage of being able to treat wet algal matter (80–90% humidity), thus freeing itself from any pre-mining drying operation. The anaerobic digestion process consists of three successive biochemical steps:

– hydrolysis of complex compounds (production of simple sugars from polysaccharides);

– fermentation (transformation of simple sugars into alcohol, acetic acid, volatile fatty acids and gas);

– methanogenesis (production of methane from fermenting gases).

However, the C/N ratio in microalgae is relatively low, as many species have high protein contents, and anaerobic digestion relies on the hydrolysis and fermentation of complex sugars. In order to improve the anaerobic process, the addition of waste paper to the algal biomass is recommended. In a 50/50 ratio, this doubles methane production to 1.17 mL/L per day, compared to 0.57 mL/L per day without the addition of paper (Brennan and Owende 2010).

Alcoholic fermentation, an ancestral biotechnological process, is also applied for the production of ethanol from algal biomass. On the species *Chlorella vulgaris*, fermentation for bioethanol production has been successfully tested. This alga proves to be a good source of ethanol due to its high starch content, and the observed conversion rate of biomass to ethanol is 65% (Brennan and Owende 2010).

Hydrogen is also an energy source that can be used as a fuel. During the photosynthetic process, hydrogenase, the enzyme that converts protons (H^+) into hydrogen (H_2), is inhibited by the production of photosynthetic oxygen. The lifting of this inhibition is possible if the algae are put in an anaerobic condition. This principle is at the basis of the third process of biochemical conversion or photobiological production of hydrogen. The most commonly used technique separates the biomass growth stage under photosynthetic conditions from the stage where the algae will end up in anaerobic conditions and produce hydrogen. According to some authors, this sequential sequence of "photosynthesis-hydrogen production" steps could theoretically produce 20 g H_2/m^2/day (Melis and Happe 2001).

Among the notion of biofuel (liquid or gaseous), there is a particular biofuel qualified as biodiesel. Biodiesel can be made from a mixture of oilseeds and algal biomass that can be used directly in diesel engines. It is composed of algal oil made up of long-chain fatty acids. After extraction, this oil undergoes a transesterification reaction between fatty acids and alcohol, generating mono-esters qualified as biodiesel.

The production of biodiesel from algal biomass is based on a first step which is the realization of an "algae cake" obtained after filtration and drying of the culture. This algae concentrate represents 15–25% of the solid matter in suspension. Oil extraction is carried out from this concentrate and the esterification reaction between fatty acids (triglycerides) and alcohol (methanol) will be carried out in a molar ratio of 6:1 (Mata *et al.* 2010). The conversion efficiency of the process makes it possible to establish that one liter of oil generates almost one liter of biodiesel. The production of biofuel requires the selection of species whose biochemical composition is best suited to the type of product desired.

4.5.2. *Algal species used as biosources*

The lipid, polysaccharide and protein fractions of microalgae can be converted into fuel. The lipid fraction of microalgae consists mainly of hydrocarbons quite similar to those found in petroleum. This fraction can be easily converted to biodiesel (see section 4.5.3). Species belonging to the genus *Chlorella* have lipid contents ranging from 14 to 63% of dry matter, depending on growing conditions and interspecific variations. The *Botryococcus braunii* species can have a lipid content of up to 86% of dry matter (see Table 4.9).

The polysaccharide fraction can be transformed into bioethanol by alcoholic fermentation. Algae such as *Porphyridium cruentum* with a polysaccharide content of up to 57% of the dry mass are therefore suitable for this kind of bioconversion.

Species	**Lipid content (% MS)**
Botryococcus braunii	25–86
Chlorella emersonii	25–63
Chlorella protothecoides	14–57
Chlorella vulgaris	14–56
Schizochytrium sp.	50–77
Neochloris oleoabundans	35–65

Table 4.9. *Examples of lipid contents of some microalgae species potentially recoverable as biodiesel (from Ghasemi* et al. *(2012))*

The protein fraction appears less interesting than the other two for conversion into biofuel because it is less rich in carbon and hydrogen. This implies that species with high protein content, such as spirulina, are less interesting candidates for biofuel production. However, all three fractions, including protein, can be converted into biomethane during the anaerobic digestion process.

4.5.3. *The economic context*

The production of fuel oil from algal biomass is an interesting environmental issue because it generates non-toxic products using a renewable biological resource. However, the economic feasibility of this approach depends on the comparison of the costs of production and treatment of biomass with the cost of production of fuel oil from fossil fuels.

A study showed that the costs of algal oil production were strongly influenced by the techniques used upstream for the cultivation of microalgae. Indeed, the production cost of a liter of biodiesel obtained from an open-pond culture is $2.18 per liter, compared to $4.56 per liter for biomass grown in a photobioreactor (see Table 4.10) (Davis *et al.* 2011). As an indication, the production cost of a liter of diesel averages $0.57.

Growing method	Production cost Liter oil	Production cost Liter biodiesel	Production cost Liter diesel
Open basin	$1.89	$2.18	$0.57
Tubular photoreactor	$4.02	$4.56	

Table 4.10. *Comparison of production costs between biofuel and diesel (based on Davis* et al. *(2011))*

This economic comparison, although not very favorable to the production of biodiesel from microalgae, is not fatal to the development of this sector. Indeed, at a time of depletion of fossil fuels and climate change, the production of biofuels from a resource capable of fixing CO_2 and producing oxygen remains a significant societal and economic challenge.

Thirteen major companies, including the European company Nestlé, are therefore developing this new process for converting microalgae into biofuel. Of these, eight are American; the others are mainly European (see Figure 4.9). Eleven out of thirteen companies produce in open systems, and two have chosen to produce in closed systems (Singh and Gu 2010).

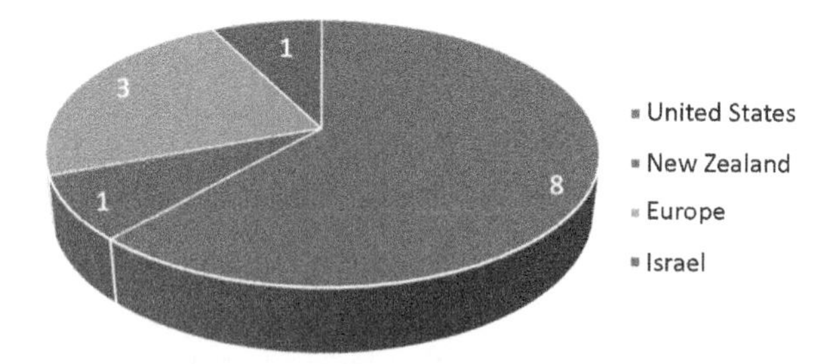

Figure 4.9. *Distribution of industrial companies producing biofuel from algal biomass conversion (based on Singh and Gu (2010)). For a color version of this figure, see www.iste.co.uk/fleurence/microalgae*

4.6. Other applications

In addition to the classical compounds previously mentioned, there is a category of molecules belonging to the family of biotoxins that is also being developed. These toxins are mainly produced by microalgae belonging to the dinoflagellate group. Their toxic properties are of interest to medical research and the pharmaceutical industry. Okadaic acid (see Figure 1.7), a well-known marine toxin, has been used in medical cancer research. Its role in inducing angiogenesis and promoting metastatic proliferation has been particularly studied (Kim *et al.* 2009). This toxin is believed to act on the activity of the "hypoxic inducible factor-1" (F1-IH), which is a growth factor of the vascular endothelium. The cytotoxic activity of okadaic acid against murine and human tumor cell lines has also been tested (see Table 4.11) (Islam *et al.* 2002). The highest inhibitory activity was observed for the human stomach cancer MKN 7 line (see Table 4.11).

Tumor cell line	CI50 okadaic acid (10^{-9} M)
HeLa human brain cancer	10.65
MKN 7 human gastric cancer	9.50
Murine colon adenocarcinoma (clone 26)	13.50
Murine sarcoma (S180)	12.70

Table 4.11. *Cell growth inhibition activity of human and murine tumor lines by okanoic acid (expressed as Inhibitor concentration of 50% of growth) (according to Islam* et al. *(2002))*

On the contrary, *per os* administration of okadaic acid to mice over 24, 36, and 48 hours did not for concluding on the effects of this toxin on digestive tissues (intestine and colon) (Morito *et al.* 2011).

In vitro, okadaic acid also proves to be a powerful inhibitor of phosphatase type 1 (PP1) and 2 A (PP2A), enzymes that play a fundamental role in protein phosphorylation (Fernandez *et al.* 2002). This toxin is therefore often used in research when inhibition of these phosphatases is required.

In addition to okanoic acid, there is another toxin whose activity is of interest to medical research and the pharmaceutical industry. This is tetrodotoxin (see Figure 4.10). It is synthesized, among others, by the dinoflagellate, *Alexandrium tamarense* (Kodama *et al.* 1996), and is the subject of a pharmaceutical valorization trial. This toxin gives rise to the development of a product known as Tetrodin™. This molecule is being developed by WEX Pharmaceutical Inc. as a pain treatment in cancer therapy (Gallardo-Rodriguez *et al.* 2012). It possesses analgesic properties based on the inhibition of nerve impulses traveling through the peripheral nervous system, thereby limiting pain. This molecule is presented as an alternative to opiates, whose tolerance is very reduced in cancer patients with comorbidity factors (Hagen *et al.* 2017). Its prolonged analgesic effect over several weeks, only four days after administration, seems to be the main advantage of this molecule. Despite its pain-relieving effects and the completion of all phases of the clinical study (I, II, III), this drug is still not produced (personal communication, Grovel O.).

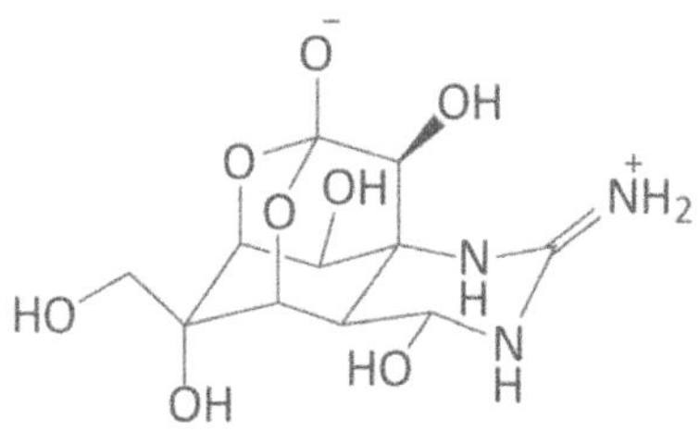

Figure 4.10. *Structure of tetrodotoxin (source: Grovel O.)*

In a more anecdotic way, tetrodotoxin, a powerful neurotoxin, also seems to be used in Voodoo ceremonies and is said to be at the origin of the folklore of zombies or the living dead. The Haitian penal code recognizes the existence of a poison at the origin of this phenomenon since it stipulates:

> It is also qualified as an attempt on a person's life the use against him of substances which, without causing death, produce a lethargic effect that is more or less prolonged, in whatever way these substances have been administered, whatever the consequences may have been. If as a result of this lethargic state the person has been buried, the attack will be qualified as murder (Davis 1983).

The analysis of the "zombie" drug revealed the presence of tetrodotoxin, and the lethargic effects of this neurotoxin are related to that of death. The provision of an antidote by the sorcerer could explain the "so-called return to life".

This more than comical application of tetrodotoxin is nonetheless an ethnobotanical valorization linked to human beliefs.

Numerous patents on the applications of dinoflagellate toxins have been filed in recent years; however, they mainly concern methods for detecting these toxins in the environment. Conversely, patents on the application of these toxins as new drugs or products for industry are very rare.

5

Extraction Processes

The valorization of molecules or fractions of microalgae passes through by the implementation of efficient and, if possible, non-denaturing extraction processes in order to preserve the physico-chemical and biological properties of the substances to be recovered. These processes can be based on physical mechanisms such as grinding, ultrasonic extraction, high-pressure extraction, super-critical CO_2 fluid or microwaves. These methods can also be used in combination. Although most of these methods are recent, they belong to classical processes, as they are frequently used. Other more recent and less used processes are also possible; these are more particularly methods of enzymatic engineering. These methods are based on the use of enzymes, generally, polysaccharidases, which, by digesting the wall of microalgae, will promote the extraction of intracellular compounds. Finally, alternative processes such as extraction by electroporation or pulsed electric fields are also possible depending on the nature of the molecules to be extracted (e.g. RNA).

5.1. Conventional processes

5.1.1. *Ball mills*

The grinding of algal biomass remains the oldest and easiest method to implement. It is the mechanical method *par excellence*. However, this method has been improved in order to avoid denaturation phenomena of the released intracellular substances. In particular, a process involving the grinding of algal biomass by collision with beads has been developed. It was successfully tested on the species *Chlorella vulgaris* (Postma *et al.* 2015). In this concept, the algal suspension is agitated in a grinding chamber in the presence of moving beads that will disintegrate the microalgae. The stirring speed in the extraction medium is between 6 and 12 m/s. This parameter is fundamental, as it has a direct influence on the efficiency of the grinding process. Indeed, the optimal efficiency for cell disintegration and protein solubilization is obtained for a speed of 9–10 m/s, regardless of the concentration of biomass introduced into the chamber (Postma *et al.* 2015). In order to avoid thermal denaturation of the released substances due to heating after grinding, the temperature in the chamber is regulated by a cooling coil. This system keeps the temperature below 35°C. For this reason, the process is referred to as a "gentle grinding method". As with any grinding operation, the efficiency of the process also depends on the application time. In the case of *C. vulgaris*, three parameters have been studied in particular as a function of time, namely:

– the rate of cellular disintegration;

– the level of solubilized proteins;

– the rate of pigments released.

The optimal cell disintegration rate (>95%) is observed after 250 seconds. From 200 seconds onwards, 90% of the proteins were solubilized. The maximum concentration of protein thus extracted is 17.5 mg/mL. In contrast, just under 75% of the pigments are extracted under these conditions.

However, the conditions for bead grinding differ depending on the species. Thus, the parameters applied to the species *Nannochloropsis oculata* and *Porphyridium cruentum* are very different (Montalescot *et al.* 2015). Indeed, at equivalent pressure, the mechanical energy required for cell disintegration of *N. oculata* is 3.5 times higher than that required for *P. cruentum*. This observation highlights the fact that the bead-grinding process must be optimized for each species. The nature of the wall, which differs according to the species, is probably at the origin of this observation.

More generally, the main advantages of extraction by ball milling are that it can be applied at high biomass concentrations (see Table 5.1) and is effective in cell disruption. However, it requires a lot of energy when applied on a large scale (Phong *et al.* 2018).

5.1.2. *Ultrasonication*

Ultrasonication belongs to the category of physical extraction methods which are classically used. This method is based on the use of ultrasound which will cause cell disintegration. The extractive efficiency of the process depends on the frequencies used and also varies according to the species considered (Wang *et al.* 2014a). The use of a high-frequency treatment (3.2 MHz) appears to be more effective in cell disintegration than a low-frequency treatment (20 kHz). However, the combination of the two treatments induces a better efficacy in cell disruption. This has been demonstrated in particular in the species *Scenedesmus dimorphus* and *N. oculata* (Wang *et al.* 2014a, 2014b). Although an extractive method in its own right, ultrasonication is often used in combination with other methods (see Table 5.1). In particular, it can be combined with extraction using organic solvents such as methanol or heptane. In this case, it is referred to as ultrasound-assisted extraction. This type of method has been used to extract β-carotene from *Spirulina platensis* (*Arthrospira platensis*).

In this process, a pre-treatment is carried out by stirring the Spirulina biomass in methanol (2 min). Extraction is then carried out using heptane and by applying an electrical acoustic intensity of 167 W/cm^2 (Dey and Rathod 2013). The prior application of methanol treatment is essential in this process since it increases the pigment extraction yield by a factor of 12.

Apart from pigments, the extraction of lipids in the presence of organic solvents and assisted by the use of ultrasound is also a widely used method. Sonication improves the penetration of solvents into the algal biomass and thus facilitates the extractive capacity of the solvents used. This process has notably been used on a mixed culture of microalgae from treated water effluents (Keris-Sen *et al.* 2014).

This type of extraction based on the ultrasonication-assisted use of organic solvents is a process that does not fit into the concept of "green chemistry". However, ultrasound used alone can also extract apolar compounds such as lipids. In this particular case, ultrasonication, therefore, appears to be a "green" process with a very limited environmental impact (see Table 5.1). This process has been applied for the extraction of lipids from the microalga *N. oculata* (Adam *et al.* 2012). It is applied from a fresh seaweed paste (70% humidity) and with a low ultrasound frequency (20 kHz). The recovery yield of algal oil from this process is 0.2%, while it is estimated at 5.7% by means of conventional extraction using organic solvents (chloroform/methanol). However, this disadvantage can be circumvented by the ease of treating algal biomass on an industrial scale using ultrasound. Indeed, there are ultrasonic reactors capable of extracting 10, 20 or 200 kg of algae per hour (Adam *et al.* 2012).

Extraction method	Mechanism	Advantages	Disadvantages
Ball grinding	Physical collision with the balls	– Treatment of a concentrated biomass – High efficiency of cell disintegration – Prevention of overheating phenomena by cooling coil	– Large-scale energy expenditure on a large scale – Variable effectiveness depending on the species
Ultrasonication	Shear forces by cavitation	– Less species-dependent universal process – Easily combinable with other extraction methods – Reduced need for energy and consumption of solvents – Limited environmental impact – Easily transposable to pilot scale – Low cost of required facilities	– Implementation of systems to prevent warming
Enzymatic hydrolysis	Substrate degradation	– Very specific process – Mild operating conditions – Easily transposable to pilot scale	– Expensive and time-consuming process – Knowledge of the nature of the walls
Pulsed electric fields	Membrane destabilization	– Low environmental impact	– Difficult transition to pilot scale

Table 5.1. *Advantages and disadvantages of some extraction methods applied to microalgae (based on Phong* et al. *(2018))*

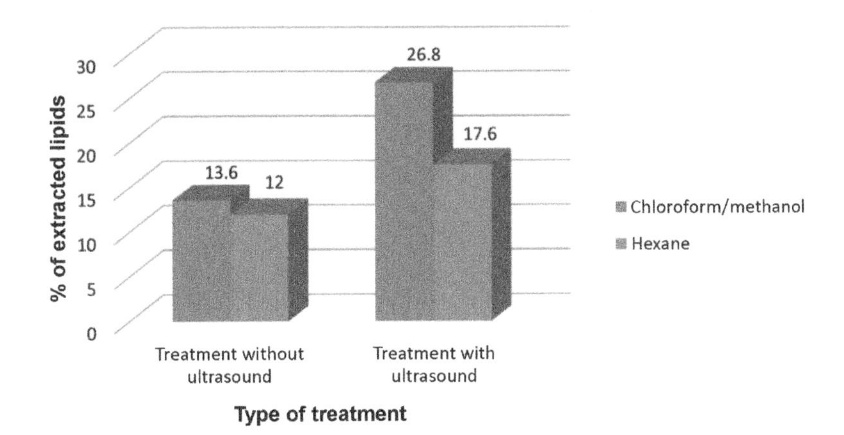

Figure 5.1. *Influence of ultrasonication on lipid extraction using organic solvents from a mixed culture of microalgae (from Keris-Sen* et al. *(2014)). For a color version of this figure, see www.iste.co.uk/fleurence/microalgae*

Ultrasound-assisted extraction also applies when using aqueous solvents such as ionic buffers. This methodology is used to facilitate the solubilization of polar molecules such as proteins. It has been applied to the microalgae *Chlorella vulgaris* and has the advantage of operating from wet biomass, which cancels out the costs associated with dehydration of the algal resource (Lee *et al.* 2017). The protein solubilization rate increases significantly with the power of the sonication (200–400 W) and the duration of the sonication (0–30 seconds). It is optimal for an applied power of 400 W and a duration of 30 seconds. Under these conditions, the rate of recovered proteins is around 27% of the dry matter. This represents nearly 95% of the protein initially present in the biomass.

The use of ultrasonication is not limited to the assistance of solvent extraction, whether organic or aqueous in nature. It can also be associated with other methods, such as enzymatic hydrolysis, in the extraction of carotenoid pigments from the microalgae *C. vulgaris* (Deenu *et al.* 2013) (see section 5.2).

5.1.3. *Extraction using supercritical fluid*

Supercritical CO_2 is the main fluid used in this extraction methodology. In this process, CO_2 is used above its critical physical temperature (31°C) and pressure (74 bar) points. The fluid generated then acts as an extraction solvent with low toxicity and especially chemical inertness. It is often presented as an alternative to the use of organic solvents. In addition to these advantages, supercritical CO_2 has the disadvantage of having a rather limited dissolution capacity, as it is not very polar. To overcome this, the CO_2 fluid can be combined with alcoholic cosolvents such as methanol or ethanol. The use of supercritical CO_2 is highly developed for the extraction of pigments and lipids.

In particular, it has been used for the extraction of carotenoids and lipids from the microalgae *Chlorella vulgaris* (Mendes *et al.* 1995). The process was applied from a freeze-dried biomass that had been subjected to grinding or not. In this experiment, the influence of the temperature of the CO_2 fluid and that of the pressure exerted was also the subject of particular attention. On the unground biomass, it was found that at constant pressure (35 MPa), the amount of algal oil extracted increases with the temperature applied (40–55°C). Similarly, at a constant temperature, the quantity of solubilized lipids increases with increasing pressure (20–35 MPa). Under these conditions, the highest content of extracted lipids is approximately 240 mg (48 mg/g MS or 4.8%) for a temperature of 55°C and a pressure of 35 MPa. A similar finding is observed for the extraction of carotenoids. At a temperature of 55°C, the maximum pigment content is obtained at a pressure of 35 MPa (see Figure 5.2).

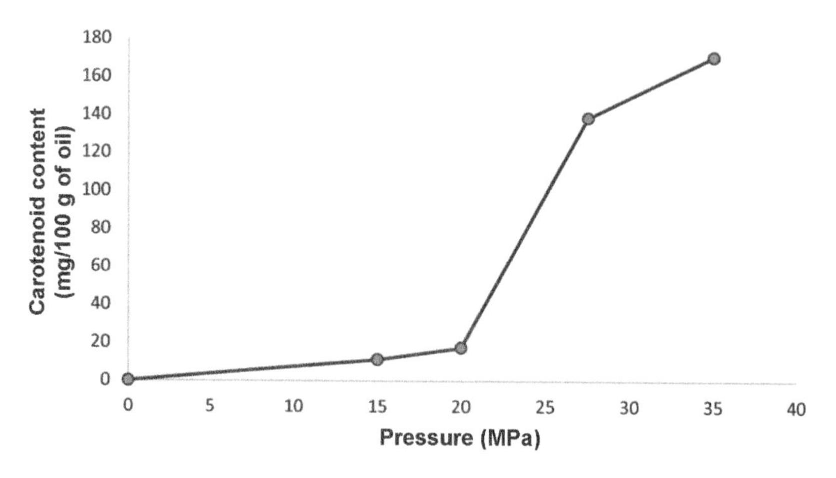

Figure 5.2. *Influence of pressure at 55°C on the extraction of carotenoids from the microalga* Chlorella vulgaris *without grinding by supercritical CO_2 fluid (from* Mendès et al. *(1995)). For a color version of this figure, see www.iste.co.uk/fleurence/microalgae*

The application of this process under optimal conditions (55°C/35 MPa) on crushed seaweed results in a lipid recovery yield of 13.3%, compared to just under 5% on uncrushed seaweed. The grinding operation, therefore, potentiates the efficiency, in this case, of supercritical fluid extraction. However, this improved efficiency rate is lower than those observed when using organic solvents such as acetone or hexane (16.8 and 17.5%, respectively).

Moreover, after grinding, the carotenoid extraction yield under the above-mentioned optimal conditions is 0.05% (expressed in relation to dry matter). This yield is 0.03 and 0.04% via extraction with hexane and acetone, respectively (Mendès *et al.* 1995). This process is, therefore, comparable in terms of efficiency to conventional processes for the recovery of carotenoids from the microalgae *C. vulgaris*.

Supercritical CO_2 fluid extraction is a method often used for the solubilization of compounds of pharmaceutical interest because it is non-toxic, and the substances can be easily purified by HPLC chromatography (Kitada *et al.* 2009). In particular, this method has been successfully tested on the microalga *Chlorella vulgaris* for the extraction of lutein and other pigments (β-carotene, chlorophyll b). Optimal extraction conditions are observed at a critical temperature of 80°C and a pressure of 50 MPa with the use of ethanol as a co-solvent (Kitada *et al.* 2009).

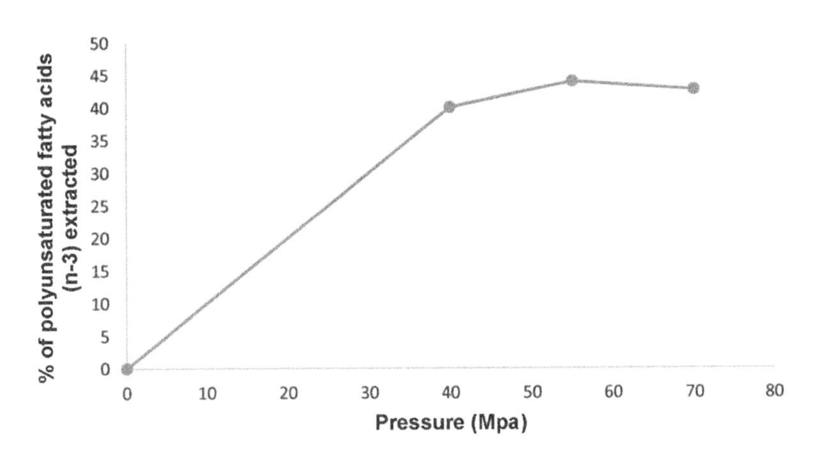

Figure 5.3. *Influence of pressure at 55°C on the extraction of polyunsaturated fatty acids from the microalga* Nannochloropsis *sp. by supercritical CO_2 fluid (from Andrich* et al. *(2005)). Percentages are related to the total fatty acid fraction. For a color version of this figure, see www.iste.co.uk/fleurence/microalgae*

It has also been used for the extraction of the bioactive lipid fraction of the microalga *Nannochloropsis* sp. (Andrich *et al.* 2005). For this species, the extraction of polyunsaturated fatty acids from the omega-3 series is optimal with a fluid administered at a temperature of 55°C and a pressure of 55 MPa (see Figure 5.3).

Under these conditions, the extraction of eicosapentaenoic acid is optimal (33% of the total fatty acid fraction). The extractive efficiency is notably higher than that observed during a conventional solvent extraction (31.1% in the presence of hexane) (see Figure 5.4).

The supercritical CO_2 extraction method is, therefore, an interesting method for the solubilization of molecules of interest, such as pigments or fatty acids of interest in terms of valorization in nutrition and human health. It allows these substances to be obtained under conditions that are independent of the use of toxic solvents such as chloroform or hexane. This last aspect is one of the main advantages of extraction with a supercritical fluid and explains, in particular, the success of this process on an industrial scale.

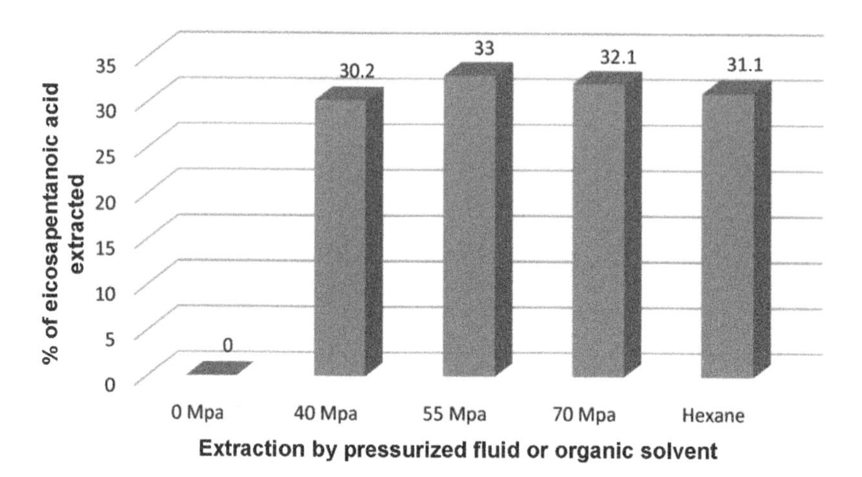

Figure 5.4. *Influence of pressure at 55°C on the extraction of eicosapentaenoic acid from the microalga Nannochloropsis sp. by supercritical CO_2 fluid and comparison with a conventional extraction with hexane (from Andrich* et al. *(2005)). Percentages are related to the total fatty acid fraction. For a color version of this figure, see www.iste.co.uk/fleurence/microalgae*

5.1.4. *Extraction by microwaves*

Extraction by means of microwaves is a classical physical process. Like ultrasonication, this physical method has a destructuring effect on the wall and membrane of microalgae. Its use also has the advantage that it can be applied to wet biomass, which considerably reduces the cost of processing the raw material by avoiding dehydration operations.

The use of microwaves has been successfully tested for the extraction of oil from the microalga *Nannochloropsis gaditana* (Menéndez *et al.* 2014). In particular, its use was compared with that of extraction using ultrasound and organic solvents (chloroform/methanol). For all the methods compared, the application time is a limiting factor in the efficiency of the extraction process. For conventional processes, based on extraction with organic solvents, the best fatty acid recovery rate (13.59% expressed on a dry matter basis) is obtained after an extraction of 45 minutes at 60°C. For comparison, at the same temperature and after 20 minutes, the microwavable extraction allows the recovery of 14.36% of fatty acids. The ultrasonic process, notably using a cavitation tube, recovers 14.76% of the same fraction. These two physical processes, although they require some methanol, avoid the use of large quantities of a chlorinated solvent such as $CHCl_3$, which is particularly toxic for the environment. For a short exposure time of 10 minutes, the method based on the use of microwaves at a temperature of 90°C and under pressure appears to be the most effective technique for recovering the fatty acid fraction of the algae (see Figure 5.5).

This is also true for the extraction of eicosapentaenoic acid, where the recovery rate (1.18% of dry matter) is optimal with the microwave method (10 minutes, 90°C).

Extraction by means of microwaves is a physical process, like other alternative processes (ultrasonication, high pressure), for the extraction of the lipid fraction of algae, using expensive organic solvents that are above all harmful to the environment and human health.

Moreover, the microwave process, like that of ultrasonication, is easily transposable to an industrial scale, which also represents one of its main assets.

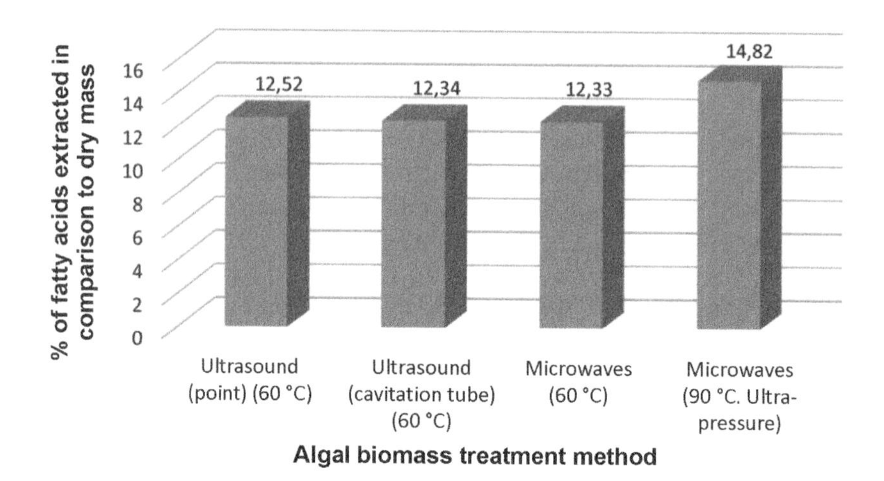

Figure 5.5. *Comparison of microwave and ultrasonic methods in fatty acid extraction from the microalga* N. gaditana *after 10 minutes of application (from Menéndez* et al. *(2014)). For a color version of this figure, see www.iste.co.uk/fleurence/microalgae*

5.1.5. *High-pressure extraction*

Among the physical methods applied to the extraction of algal metabolites, the high-pressure extraction method is an efficient and very easily usable process from an industrial point of view. This method leads to the rupture of the cell and causes the release of intracellular substances. It has been successfully applied to the biomass of *Nannochloropsis* sp., *Chlorella* sp. and *Tetraselmis suecica* (Spiden *et al.* 2013). The pressure required to achieve 50% cell disruption varies greatly depending on the species (see Figure 5.6). It is particularly high for the biomass of *Nannochloropsis* sp. since it is established at about 2,000 bars.

Extraction by means of high pressure can also be used as a selective process to solubilize certain compounds of interest, such as pigments belonging to the phycobiliproteins family. In this case, the extractive method also acts as a pre-purification method of these compounds. In the microalga *Porphyridium cruentum*, the use of different pressures has notably allowed a more selective separation between proteins and B-phycoerythrin (B-PE) (Jubeau *et al.* 2013). Thus, at pressures below 90 MPa, protein extraction is more selective. The selectivity of this extraction is lost at pressures equal to or greater than 170 MPa. The use of low pressures ($\leq$ 70 MPa) also influences the extraction of B-PE. Thus, the amount of B-PE extracted increases from

5% (expressed as a percentage of the total protein fraction) when a pressure of 27 MPa is applied to 35% at 70 MPa. With this differential pressure application system, it is possible to selectively extract the proteins in a first step where low pressures are applied and to subsequently recover the B-PE, whose maximum extraction is observed at a pressure of 270 MPa (see Figure 5.7).

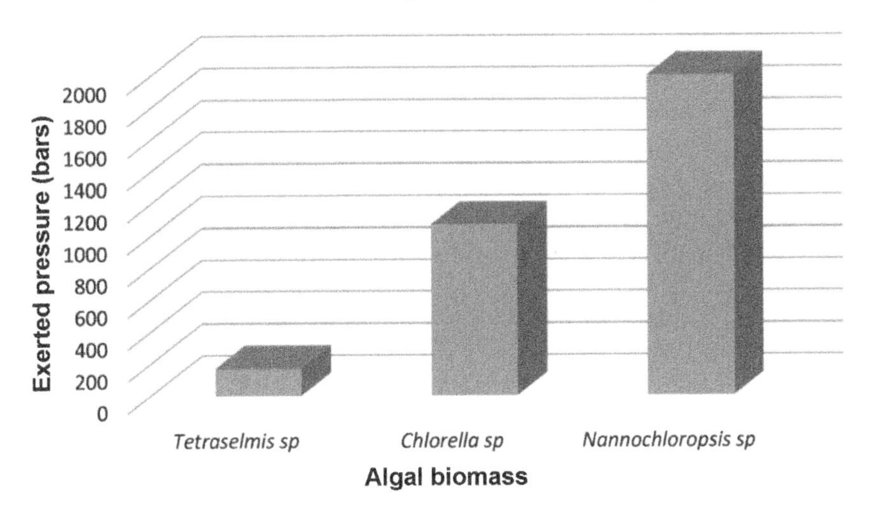

Figure 5.6. *Effect of pressure on the decay of 50% of the cells in the biomass of different species of microalgae (from Spiden* et al. *(2013)). For a color version of this figure, see www.iste.co.uk/fleurence/microalgae*

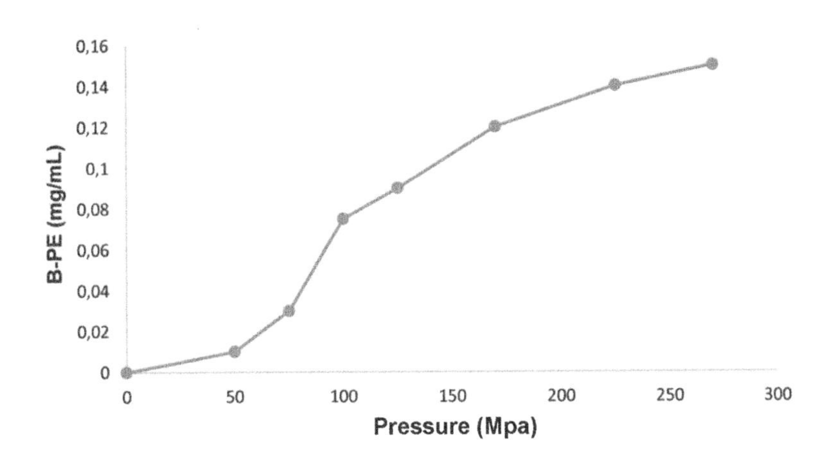

Figure 5.7. *Effect of pressure on the extraction of B-phycoerythrin (B-PE) from the microalgae* Porphyridium cruentum *(from Jubeau* et al. *(2013)). For a color version of this figure, see www.iste.co.uk/fleurence/microalgae*

High-pressure extraction is not limited to the extraction of hydrosoluble protein pigments such as B-PE. This process is also applicable to solubilization of carotenoids or chlorophylls. In particular, it has been validated for the extraction of these pigments from the alga *Chlorella vulgaris* (Cha *et al.* 2010). Extraction under high pressure is done in the presence of a methanol/water mixture (90/10). At a temperature of 160°C and for a period of 30 minutes, this method allows a significantly improved extraction of β-carotene (2–5 times) compared to conventional processes.

Extraction in liquid medium under high pressure remains a method of interest for obtaining molecules whose physico-chemical or biological properties remain intact. Its extrapolation on an industrial scale also remains a major advantage of this extraction process.

5.1.6. *Extraction facilitated by lyophilization*

Freeze-drying of algal material is sometimes used as a pre-treatment to facilitate the extraction of compounds of interest. This type of approach has notably been developed to improve the solubilization of proteins and enzymes from the microalga *Chlorella vulgaris* (Unterlander *et al.* 2017).

Freeze-drying of algal biomass, prior to the application of any subsequent grinding operation, aims at destructuring the cell wall via *the* sublimation process (freeze-dehydration). It is particularly recommended as a pre-extraction process for the treatment of algal cells with complex and highly cross-linked cell walls. This approach has demonstrated its effectiveness in the extraction of *C. vulgaris* proteins. Freeze-drying of fresh biomass has been applied in particular beforehand for the implementation of classical extraction techniques, such as mortar grinding, ball grinding, ultrasonication or pressure de-segregation (French press) Freeze-drying before applying extraction to bead milling improves protein recovery efficiency by a factor of 3 (see Table 5.2). In the case of extraction using a French press, freeze-drying the algal biomass beforehand increases the protein extractive yield by a factor of 7.5 (see Table 5.2).

Regardless of the extraction technique used, freeze-drying very significantly improves the quantity of proteins extracted from *C. vulgaris*. It also facilitates the solubility of enzymes such as glycerol kinase or pyruvate kinase (see Figure 5.8).

	French press	Grinding with mortar	Ball mill	Ultrasonication
Proteins extracted without lyophilization (mg/g MF)	8	25	25	6
Proteins extracted after lyophilization (mg/g of MF)	60	55	75	20

Table 5.2. *Effect of freeze-drying or lack of freeze-drying on protein extraction from* C. vulgaris *by different mechanical or physical techniques (from Unterlander* et al. *(2017))*

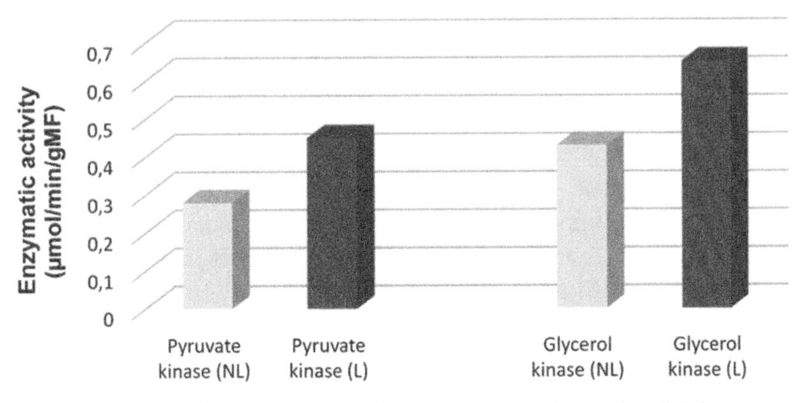

Figure 5.8. *Effect of freeze-drying on the French press extraction of kinase enzymes from microalgae* Chlorella vulgaris *(from Unterlander* et al. *(2017)). For a color version of this figure, see www.iste.co.uk/fleurence/microalgae*

Freeze-drying thus appears as a useful and auxiliary procedure to the mechanical and physical extraction techniques applicable to microalgae with a wall resistant to physical stresses. Its use on a laboratory or industrial scale does not present any particular difficulties and can therefore be justified within the framework of a valorization of the microalgal protein or enzyme fraction.

5.2. Enzymatic hydrolysis

Enzymatic hydrolysis is a process of destructuring the wall of microalgae through the use of enzymes that degrade wall compounds such as polysaccharides. This approach is an alternative to the use of conventional physical or mechanical processes. However, it requires knowledge of the biochemical composition of the wall and the availability of the appropriate enzymes. The cell wall of microalgae is generally composed of an inner and an outer layer (Demuez *et al.* 2015). The outer layer is often composed of polysaccharides such as pectin, agar, carrageenans and alginates depending on the species. The inner layer consists of cellulose microfibers, hemicellulose and glycoproteins. The composition, structure and complexity of the wall vary according to species, stage of development and season (Demuez *et al.* 2015). For example, the wall of the species *Nannochloropsis gaditana* consists of an inner layer mainly composed of cellulose (75% of the mass) and an outer layer of aliphatic polymers called algaenanes (Scholz *et al.* 2014). This hydrophobic layer of algaenanes plays a protective role and functions as a bioplastic. In the microalga *Chlamydomonas reinhardtii*, the parietal ultrastructure appears more complex. In particular, it is organized in six distinct layers containing pectin–cellulose type carbohydrate associations and other hydroxyproline-rich glycoproteins (Demuez *et al.* 2015). From a biochemical point of view, this wall is divided into two domains, one poor in proteins and the other containing about 20 proteins associated with each other by non-covalent bonds (Iman *et al.* 1985). The part containing few proteins, but mainly polysaccharides, is considered as the main structural framework of the wall.

The biochemical heterogeneity of the parietal composition of microalgae is a limiting condition for the enzymatic hydrolysis process because the latter, in order to be optimized, often requires the use of several enzymes with very different optimal conditions (optimum pH, optimal temperature).

Despite these difficulties, enzymatic wall hydrolysis is a process often used to extract or transform cellular compounds of interest from microalgae. This is typically the case when using cellulases (β-glucanases) and amylase. In this context, the objective is twofold, namely, to degrade the wall structure and to hydrolyze the parietal (cellulose) or storage (starch) polysaccharides into simple sugars that are easily fermentable. This last aspect is then similar to saccharification.

The degradation of cellulose generally occurs through the successive use of an endo- and exo-β-(1,4)-D-glucanase to degrade the parietal cellulose and then the released cellulose chains. Starch is degraded by an endo-amylase to dextrin. The latter is subsequently hydrolyzed into glucose or oligosaccharide via an amyloglucosidase.

The simple sugars extracted and resulting from the transformation thus become good substrates for the fermentation process and thus the production of bioethanol.

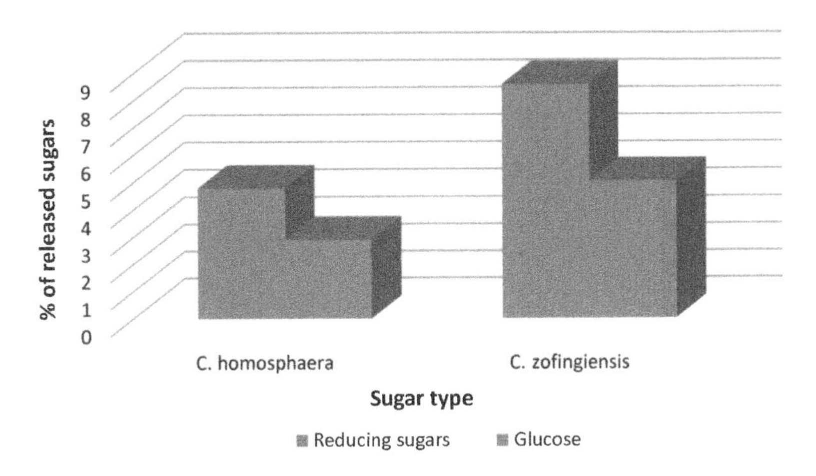

Figure 5.9. *Effect of enzymatic hydrolysis (cellulases, xylanases, amylases) on the extraction and transformation of parietal carbohydrates from the microalgae* Chlorella homosphaera *and* zofingiensis *(from Rodrigues and da Silva Bon (2011)). For a color version of this figure, see www.iste.co.uk/fleurence/microalgae*

This valorization approach, via the enzymatic treatment of algal biomass, has notably been described for Chlorella. This microalga is characterized by a cell wall composed of 80% polysaccharides, mainly cellulose (Rodrigues and da Silva Bon 2011). The biomass was treated in the presence of a mixture of enzymes consisting of cellulases, xylanases and amylases for 24 hours (pH 4.8, temperature 50°C). The saccharification yield of the process differs according to the Chlorella species considered. It is significantly lower for the *Chlorella homosphaera* species (see Figure 5.9).

However, enzymatic hydrolysis of algal biomass is not limited to the extraction of sugars useful for the production of bioethanol. This process can also be used to improve the extraction of lipids (Zuorro *et al.* 2016) (see Table 5.3), proteins (Sierra *et al.* 2017) or the production of protoplasts (Gerken *et al.* 2013).

Polysaccharidases are not the only enzymes that can be used in this type of process. Enzyme cocktails comprising glucanase and protease activities are also used for the degradation, in particular, of the wall of *Chlorella vulgaris* or *Scenedesmus dimorphus* (Demuez *et al.* 2015). These are mainly commercial cocktails such as Snailase©, which is a mixture of cellulases, amylases and proteolytic activities (see Table 5.3). Other commercial enzyme preparations, such as Celluclast© or Novozyme©, mainly with endo-β-1-4-glucanase (cellulase) activities, are also used to hydrolyze the algal cell wall and facilitate the extraction of compounds of interest.

Algal species	Enzymes	% of extracted lipids
Chlorella vulgaris	– Snailase – Trypsine	49.8
Chlorella vulgaris	Extract of *Flammeovirga yaeyamensis* (cellulases, amylases)	21.5
Chlorella vulgaris	– Celluclast 1.5 L – Novozyme 188	10
Chlorella pyrenoidosa	Immobilized cellulase	56
Nannochloropsis oculata	– Viscozyme – Proteinase K	–

Table 5.3. *Examples of commercial and non-commercial enzyme preparations for lipid extraction on selected microalgae species (from Demuez* et al. *(2015))*

Even if the use of exogenous enzymes remains the rule in the majority of cases, an alternative based on the activation of endogenous enzymes is also possible. Activation of the endogenous process of cell wall hydrolysis has the advantage of avoiding the exogenous supply of enzymes, the cost of which can sometimes be high. This approach requires a good knowledge of algal metabolism and physiology and is therefore not easily applicable to all species of interest.

The species *Chlamydomonas reinhardtii* produces under certain physiological conditions a protease that degrades the cell wall. This protease, called autolysin, is activated when the alga is grown in nitrogen-deficient environments (Sierra *et al.* 2017). This enzyme specifically hydrolyzes hydroxyproline-rich proteins present in the cell wall. Its use to facilitate the extraction of lipids and water-soluble proteins is therefore not contraindicated. This endogenous enzymatic hydrolysis process is, like any enzymatic process, temperature-dependent. The percentage of cell lysis after 2 hours of incubation is higher at 35°C compared to that observed at 25°C for the same incubation time. For longer incubation times (4 and 24 hours), the differences between the treatment at 25 and 35°C are smaller or even non-significant. The solubilization rate of proteins is significantly improved when the autolysin pre-treatment is combined with a secondary treatment, such as sonication or detergent (see Figure 5.10). Lipid extraction is also facilitated by the use of autolysin pre-treatment (1.6–1.8 times) compared to extraction without prior enzymatic hydrolysis.

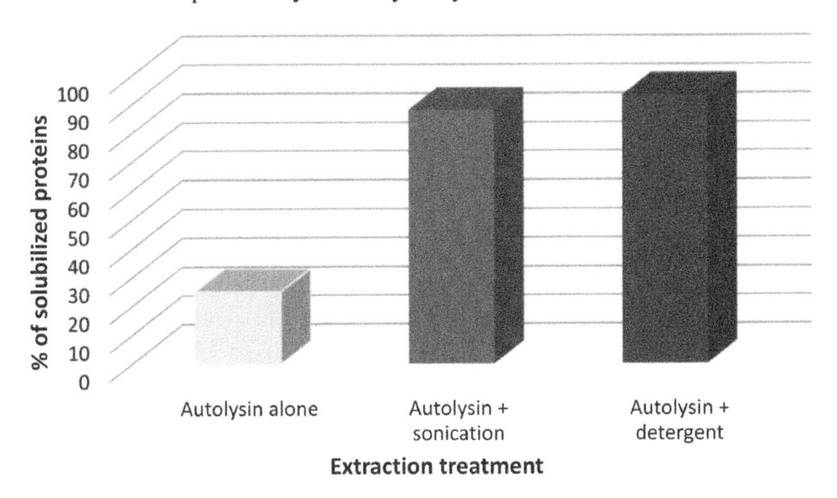

Figure 5.10. *Effect of enzymatic hydrolysis by autolysin alone or coupled with other processes on protein extraction (from Sierra* et al. *(2017)). For a color version of this figure, see www.iste.co.uk/fleurence/microalgae*

Endogenous enzymatic hydrolysis, such as that based on the use of autolysin, appears to be an effective treatment to improve the extraction of proteins and lipids from *C. reinhardtii*. It minimizes the use of mechanical grinding processes, which are more denaturing and more expensive.

The extraction process via enzymatic degradation of the cell wall is a gentle method that fits perfectly into the concept of green chemistry. This method is adapted to facilitate the extraction of compounds such as sugars, proteins, pigments, lipids or nucleic acids (DNA, RNA). It is notably used to facilitate the extraction and amplification of DNA from red algae, whether micro- or macroalgae (Fleurence and Joubert 2006). However, its effectiveness is amplified when it is associated with other methods such as sonication or the use of solvents or detergents useful for the solubilization of partially hydrophobic compounds (membrane proteins, lipids).

5.3. Other methods

Among the methods qualified by others is the method of extraction by application of pulsed electric fields. Although this is a physical method, it is very different from conventional processes based on the mechanical destruction of the wall. The principle of the pulsed-field method lies in the application to the cells of an electric field of very high intensity (2–55 kV/cm), repeatedly (pulsed) and for short periods (microsecond) in order to avoid any lethal phenomenon. Under the effect of this electric field, the cell wall and membrane will be destabilized. The membrane will be deformed, and pores will appear, allowing the exit or penetration of compounds. This principle is at the origin of the name electroporation, which is also given to this method of extraction by pulsed electric fields (see Figure 5.11).

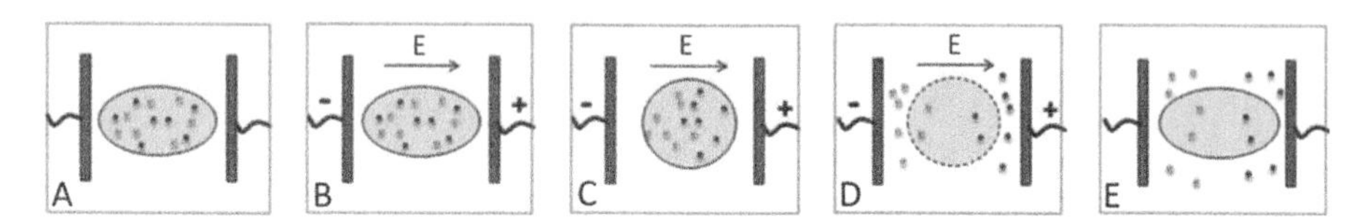

Figure 5.11. *Principle of electroporation applied to the extraction of intracellular compounds from microalgae: A (absence of electric field), B (application of an electric field), C (deformation of the cell), D (formation of membrane pores, release of intracellular compounds) and E (stop of the electric field, reformation of the cell membrane) (source: Pouchus Y.-F., from Joannes* et al. *(2015))*

Electroporation is well known in genetic engineering, where it is used to introduce a transgene into a host cell. This method, which modifies membrane permeability, is also applied as a method for extracting intracellular algae compounds. In particular, it is used to improve the extraction yield of lipids, proteins and hydrophobic pigments such as carotenoids or chlorophyll (Joannes *et al.* 2015). This method has been applied to the microalga *Chlorella vulgaris*. The application of a pulsed electric field (2.7 kV/cm) allows, in the presence of solvents, the extraction of a lipid fraction representing 22% of the fresh algal mass (Joannes *et al.* 2015). However, this methodology can lead, depending on the conditions, to the lysis of the cell. In *C. vulgaris*, lysis is observable from the application of a 4 kV/cm electric field. In *Dunaliella salina* species, cell lysis is already in place from a more moderate electric field (1.6 kV/cm). The maintenance of viability is, therefore, a function of the applied voltages and varies according to the species. The choice between viability maintenance or cell lysis is often driven by valorization imperatives, such as culture maintenance or obtaining an optimal extraction yield of intracellular compounds.

In addition to the pulsed electric field method, there is an alternative method based on the application of electric fields of lower intensity. This method uses electric fields of low voltage (10–30 V/cm) and over durations longer than a microsecond. The electric field is generated by an anode and a cathode immersed in saltwater, causing the electrolysis of the water. This process or "electrochemical method" is applied when the algal biomass is concentrated. This treatment is applied to the extraction of lipids and proteins from the microalgae *Chlorella vulgaris* (Joannes *et al.* 2015). The chemical nature of the anode plays a fundamental role in the efficiency of the lipid and protein extraction processes (see Figure 5.12(a) and (b)). The titanium/iridium oxide anode shows the best efficiency in terms of electrolysis and extractive efficiency for both proteins and lipids. This type of electrode allows the application of a much more moderate electric field (14.3 V/cm) than other types of electrodes that require the application of electric fields ranging from 26 to about 31 V/cm.

The extraction processes for molecules that can be recovered from microalgae or cyanobacteria are varied and numerous. Most of them can be used on an industrial scale, and their application depends more often on the cost of implementation in relation to the expected recovery. The optimal valorization of algal biomass is an approach that tends to reduce production and extraction costs. This is why the concept of biorefinery has become the economic principle for the valorization of microalgae in recent years.

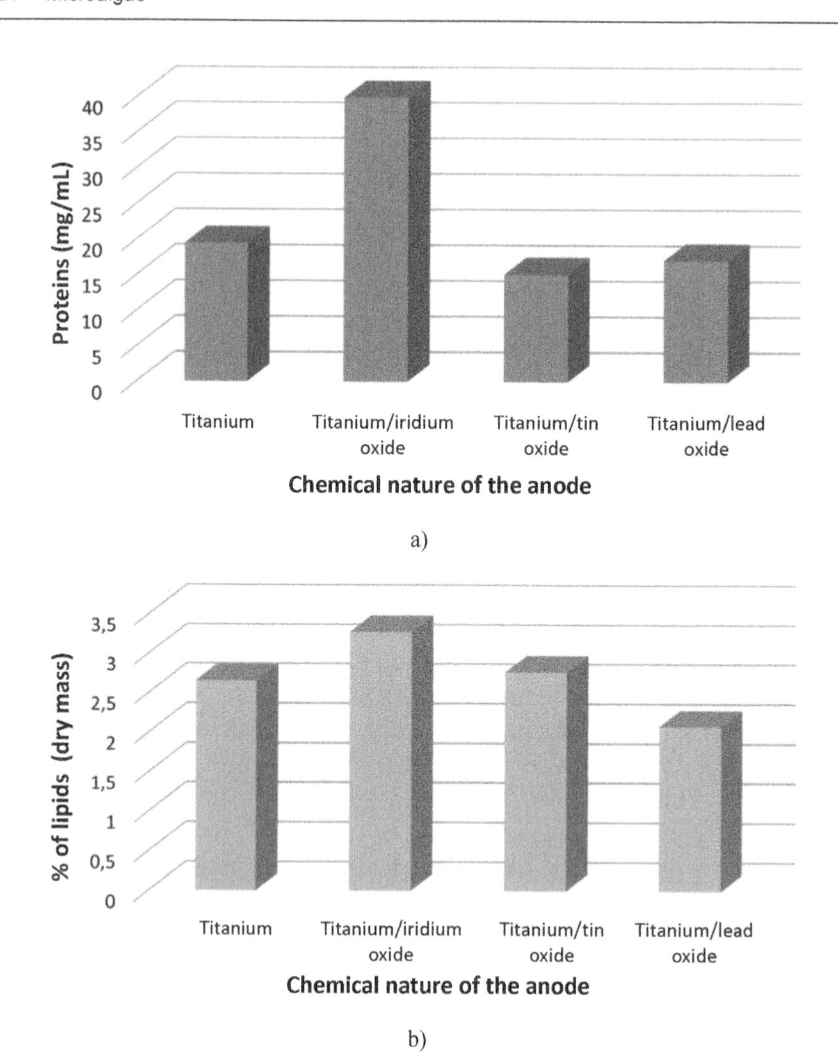

Figure 5.12. *Effect of electrochemical extraction as a function of the chemical nature of the anode on the extraction of lipids from C. vulgaris (from Joannes* et al. *(2015)). For a color version of this figure, see www.iste.co.uk/fleurence/microalgae*

6

Biotechnological Approaches

The use of microalgae and cyanobacteria as cell factories producing molecules of industrial interest is a constantly growing avenue of development. This development method is based on a concept that is becoming more and more developed: biorefinery. Its objective is to valorize all the biochemical compartments of microalgae (carbohydrates, proteins, lipids, pigments, fatty acids). It is part of a global valorization approach and not a specific approach, such as the valorization of a molecule with high added value. However, this last concept of valorization is not abandoned because there are always approaches based on the extraction of specific molecules from algae or cyanobacteria (e.g. phycocyanin, phycoerythrin).

Biorefinery and the specific production of molecules can be based on very different biotechnological processes applied to algal biomass, such as physiological forcing or genetic transformation.

6.1. Biorefinery

When microalgae are valued as food or food supplements, they are used in their entirety (see Chapter 3). The recovery of microalgae through the extraction and purification of a molecule or a class of molecules requires the implementation of purification processes that are sometimes long and costly (see Chapters 4 and 5). In this context, all other compounds or biochemical compartments of the algae are considered contaminants or waste and are therefore not recovered. Faced with this situation, the valorizers have developed a new concept, namely that of the biorefinery. This concept derives from the valorization of algae as biofuels. Indeed, algal biomass can be used in its entirety for the production of bioethanol, biogas, biomethane or hydrogen (see Chapter 4). The biochemical compartments mainly involved in the production of biofuels, namely carbohydrates and lipids, represent two of the three main biochemical categories (proteins, carbohydrates and lipids) constituting algal biomass.

Currently, the biorefinery concept has been extended to other categories of compounds or applications (see Figure 6.1). Proteins and pigments are often concerned by this extension of the biorefinery concept.

Proteins, the third biochemical compartment next to lipids and carbohydrates, are valued for animal feed and human nutrition. Meanwhile, pigments are used in many applications, ranging from human nutrition to cosmetics (see Figure 6.1).

In its concept, biorefinery aims to valorize all the biochemical compartments (proteins, carbohydrates, lipids and pigments) of algal biomass. It is mainly based on the optimization of extraction of recoverable compounds. The extracted compounds are often lipids (biodiesel) and pigments. In this case, the remaining algal biomass that still contains carbohydrates can be subjected to a fermentation process to produce biogas, such as hydrogen.

This coupling of biorefinery and gas production enhances the efficiency of the process of enhancing algal biomass. Such an approach has been successfully tested on the microalga *Nannochloropsis* sp. This species is known to have a high content of lipids and pigments, mainly carotenoids (Nobre *et al.* 2013). The biorefinery process applies to the extraction of lipids and pigments using supercritical CO_2 fluid. This extractive process allows the recovery of 85% of lipids and 70% of pigments (Nobre *et al.* 2013).

The algal biomass remaining after extraction is then subjected to a fermentation process, via *Enterobacter aerogenes*, for the generation of hydrogen with a production yield of nearly 60 mL of gas per gram of dry matter.

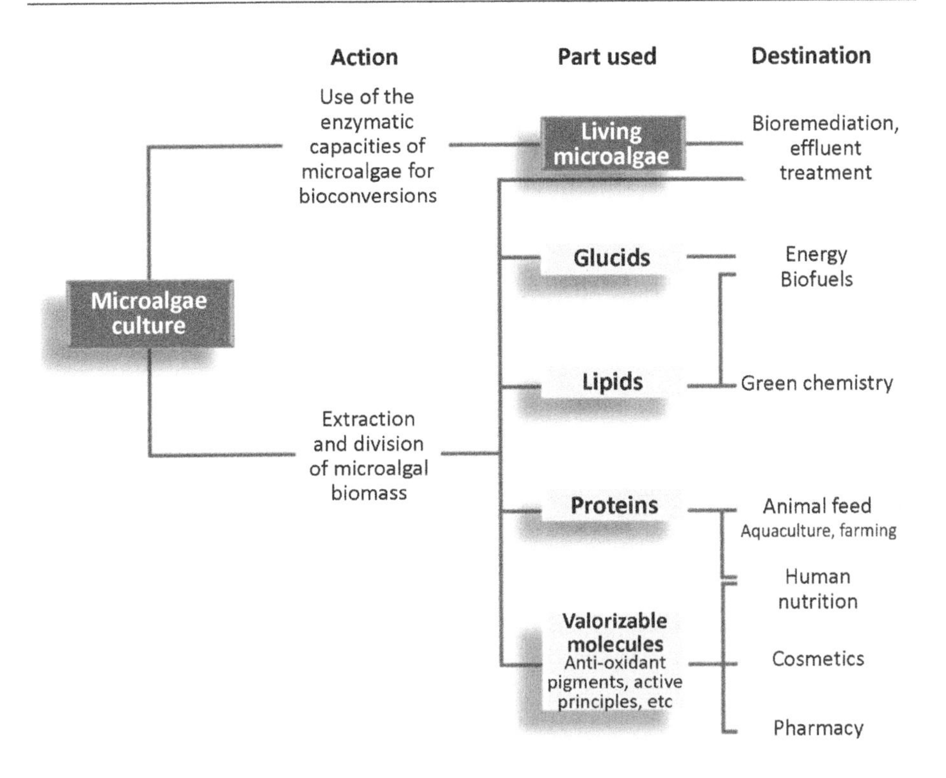

Figure 6.1. *General principle of biorefinery applied to the valorization of algal biomass (source: Pouchus Y.-F., from Wolff (2012)). For a color version of this figure, see www.iste.co.uk/fleurence/microalgae*

The efficiency of biorefinery is closely linked to the biotechnological processes applied in the production of algal biomass. Physiological forcing is one of the two main processes currently employed.

6.2. Physiological forcing

Physiological forcing is a traditional approach based on the metabolic plasticity of microalgae. It is based on the modification of culture conditions to promote the production of biomass or the activation of a metabolic pathway. In the latter case, physiological forcing is an easy way to induce the production of larger quantities of molecules of interest (lipids, pigments and enzymes). Metabolic plasticity is the result of the natural physiological plasticity of microalgae which are able to adapt to different living and growing conditions. This property has been used to achieve physiological forcing through different production modes.

Microalgae can be effectively cultivated through three different mechanisms:

– the phototrophic or autotrophic mode;

– the heterotrophic mode;

– the mixotrophic mode.

	Photosynthesis	**Adding a source of organic carbon for growth**
Phototrophic mode	+	–
Heterotrophic mode	–	+
Mixotrophic mode	+	+

Table 6.1. *Main characteristics of the modes of production of microalgae susceptible to physiological forcing*

In the phototrophic mode, microalgae perform photosynthesis from incident light and use CO_2 as a source of inorganic carbon (see Table 6.1). The production of biomass and certain metabolites depends essentially on the lighting conditions. Thus, the production of fatty acids is optimal when the crop is subjected to green light (λ 525–565 nm) (Moreno-Garcia *et al.* 2017). Under exposure of the culture to blue light (λ 450–500 nm), an optimal effect on biomass production, and consequently on the amount of fatty acids produced, has also been reported (Moreno-Garcia *et al.* 2017).

In the heterotrophic mode, microalgae are grown in the absence of light but in the presence of organic carbon (see Table 6.1). The carbon thus provided is a source of energy that will be used for the growth of the microalgae and the production of usable molecules, particularly for the production of biofuel. The organic carbon source used is generally inexpensive (glucose, glycerol and acetate), which makes this heterotrophic production method economically interesting.

The culture in the mixotrophic mode appears as an intermediate channel between the two previously described modes. This mode has the advantage of improving growth rates compared to the other two. This is mainly due to the realization of photosynthesis and the provision of a source of organic carbon, which jointly contribute to the development of biomass.

The addition of organic carbon is a phase that is common to both mixotrophic and heterotrophic modes (see Table 6.1). Glucose is one of the preferred sources of carbon. Its use influences growth and also the production of lipids. High glucose concentrations are favorable to biomass development but unfavorable to lipid production under mixotrophic conditions. In the alga *Chlorella* sp., the addition of low concentrations of glucose to the culture medium limits cell growth but increases lipid production by 52% compared to that reported for algae grown in the presence of high glucose concentrations.

Light also plays an important role in the mixotrophic mode. Like organic carbon, this factor strongly influences the production of algal biomass. An increase of more than 320% in biomass productivity has been reported for the species *Chlorella sorokiniana*, when its culture is subjected to a continuous increase in light intensity (Moreno-Garcia *et al.* 2017).

Physiological forcing based on glucose intake and light intensity as regulatory factors is particularly effective in the case of the mixotrophic production method. It allows the latter to clearly distinguish itself from the other two modes in terms of biomass productivity and lipid accumulation (see Table 6.2). This physiological forcing, which can also occur in two sequential phases, one of increased biomass production and the other of lipid accumulation, is an asset for the mixotrophic production mode.

	Phototrophic mode	**Heterotrophic mode**	**Mixotrophic mode**
Algal biomass productivity (mg/L per day)	142.1	249.7	455.5
Total lipid productivity (mg/L per day)	47.2	67.0	111.9

Table 6.2. *Effect of production patterns on biomass productivity of the species* Chlorella sorokiniana *and the production of lipids (according to Kumar* et al. *(2014))*

Physiological forcing is also applicable to the phototrophic production mode. In this mode, algae use atmospheric CO_2 to essentially carry out their growth. Numerous studies have shown that the injection of additional CO_2 into the cultures influences the accumulation of lipids. This has been particularly demonstrated in the microalgae *Chaetoceros muelleri* (Wang *et al.* 2014b) and *Chlamydomonas* sp. (Nakanishi *et al.* 2014). In the green alga *Chlamydomonas* sp. the exposure of the culture to a 4% CO_2 supplement resulted in lipid productivity of 169 mg/L day after seven days of culture (Nakanishi *et al.* 2014). This additional carbon dioxide addition also has an effect on biomass production and total lipid content, which reach values of 3.5 g/L and 33.1% (expressed in relation to dehydrated cell matter), respectively.

In the marine diatom *Chaetoceros muelleri*, an identical finding was made. The injection of additional carbon dioxide gas into the culture (10–30%) significantly improves lipid production compared to that observed with atmospheric conditions (0.03%). Biomass productivity and lipid content are maximal for the 10% CO_2 concentration (see Figure 6.2). CO_2 thus appears as a deterrent factor for the production of lipids for valorization as a biofuel.

Carbon dioxide is not the only factor that can influence the production of lipids by microalgae. In the oleaginous species *Ankistrodesmus falcatus*, the combined effect of nitrogen, phosphorus and iron on lipid and fatty acid production has been studied (Singh *et al.* 2015). The highest lipid levels (59.65% relative to dehydrated cell mass) are observed when the crop is subjected to moderate nitrogen (750 mg/L), high iron (9 mg/L) and deficient phosphorus (0 mg/L). This type of addition also results in an increase of almost 39% in the saturated fatty acid composition of the lipid fraction compared to normal conditions of supply of these elements in the culture medium.

Other examples of physiological forcing in a phototrophic mode based on the variation of the oxygenation rate of the culture in order to induce SOD synthesis or on irradiance in order to activate the pigment synthesis with high added value (phycoerythrin, phycocyanin) have also been reported (personal communication, Gudin C.).

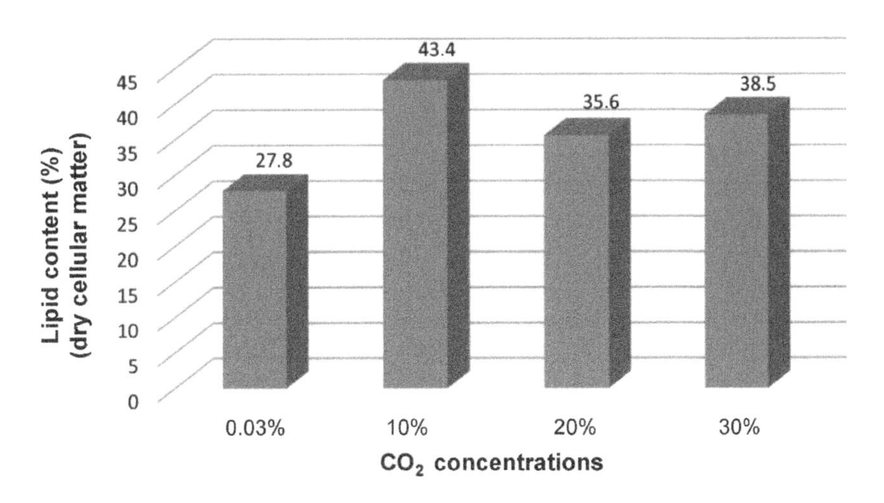

Figure 6.2. *Effect of CO_2 content on lipid accumulation in the diatom* Chaetoceros muelleri *(from Wang* et al. *(2014 a, 2014b)). For a color version of this figure, see www.iste.co.uk/fleurence/microalgae*

Among the multitude of examples, physiological forcing by concentration of the culture is a rather peculiar example, but not without interest. The principle of the method lies in cultivating algae in culture volumes less than 1/5 of the usual production volume of algal biomass (Dupré and Grizeau 1994).

Physiological forcing is a biotechnological process adapted to the three modes of production of microalgae. It is based on the metabolic plasticity of algae under environmental conditions and meets the expectations of new uses, particularly those related to the production of biofuels.

In addition to this process, which does not modify the genome of microalgae, there is another biotechnological approach based on the genetic transformation of the algae in order to make them produce substitutes of interest. This last approach is part of the more global concept of the "cell factory" already described.

6.3. Genetic transformation

Microalgae are mainly used in animal and human food or as sources of molecules of interest that they produce via their natural metabolism. In the latter case, they are considered as cell factories. However, these cell factories can be genetically transformed to produce molecules of commercial or therapeutic interest, such as proteins (Cadoret *et al.* 2008).

Unlike prokaryotic organisms (bacteria, cyanobacteria), eukaryotic organisms (terrestrial plant cells, animal cells, yeasts) are capable of assembling complex polypeptides and performing post-translational modifications such as phosphorylation or glycosylation. This last modification or N-glycosylation intervenes in the final conformation of the protein and thus has an impact on the biological activity of the latter. It also plays a fundamental role in the externalization process of endogenous proteins or in intermolecular recognition mechanisms (e.g. antigen/antibody binding).

Post-translational glycosylation is often considered a lock for the production of recombinant proteins by microalgae. This is mainly due to a relative lack of knowledge of this pathway in most species (Cadoret *et al.* 2008). Recent work carried out on the species *Chlamydomonas reinhardtii* has shown the existence of such a pathway (Mathieu-Rivet *et al.* 2013; Lucas *et al.* 2020). This species is notably capable of glycosylating proteins via the addition of xylose, thanks to the action of an enzyme or xylosyltransferase A (Lucas *et al.* 2020).

Chlamydomonas reinhardtii is a model species that has been used since the 1980s to develop transgenesis in microalgae. Its genetic transformation by the introduction of the VP1 gene of the foot and mouth disease virus enabled the production of the first vaccine protein derived from genetic engineering applied to microalgae. This genetically modified species has also been successfully used for the production of antibodies against the glycoprotein D of the herpes simplex virus (Mayfield *et al.* 2003). The production of this recombinant protein was achieved by introducing the transgene into the chloroplast genome.

Apart from *C. reinhardtii*, other species of microalgae belonging to different groups are also transformed using classical transgenesis techniques (see Table 6.3). It should be noted that most of the genetically modified species belong to the group of green algae and diatoms and that biolistics is the main method used for the introduction of the transgene.

In a fairly general way, genetic transformation is carried out at the level of the nuclear genome. It can also be carried out at the chloroplastic genome level, as is the case for the species *Porphyridium cruentum*, *C. reinhardtii* and *Euglena gracilis* (Cadoret *et al.* 2008).

As shown in Table 6.3, *C. reinhardtii* appears to be a flagship species in the development of transgenesis on microalgae, given the number of genetic transformation methods and genomic targets associated with it.

Although the production of recombinant proteins for therapeutic purposes appears to be a major stake in the development of transgenesis in microalgae, it remains that other avenues of development can justify this biotechnological approach.

Algal species	Gene transfer techniques	Genomic targets
Chlamydomonas reinhardtii **(green algae)**	– Biolistics – Electroporation – Carbide whiskers of silicon	– Nuclear genome – Chloroplastic genome – Mitochondrial genome
Dunaliella salina **(green algae)**	Electroporation	Nuclear genome
Chlorella vulgaris **(green algae)**	Electroporation	Nuclear genome
Haematococcus pluvialis **(green algae)**	Biolistics	Nuclear genome
Cyclotella cryptica **(diatom)**	Biolistics	Nuclear genome
Phaeodactylum tricornutum **(diatom)**	Biolistics	Nuclear genome
Navicula saprophila **(diatom)**	Biolistics	Nuclear genome
Porphyridium cruentum **(red algae)**	Biolistics	Chloroplastic genome
Euglena gracilis **(euglenobiontes)**	Biolistics	Chloroplastic genome

Table 6.3. *Examples of genetically transformed microalgae and the techniques used and genomic targets involved (based on Dunahay* et al. *(1995), Mayfield* et al. *(2003) and Cadoret* et al. *(2008))*

The use of genetic engineering on microalgae to increase biofuel production by modifying carbon metabolism is a valorization strategy under development (Radokovits *et al.* 2010). It is based on the modified production of carbohydrates which are metabolic precursors of biofuel, lipids and fatty acids. The transfer of genes coding for enzymes (ADP glucose pyrophosphorylase) involved in the metabolism of starch and promoting its storage is particularly targeted from the model species *C. reinhardtii*.

Knowledge of the metabolic pathways of algae is therefore necessary for the implementation of a genetic transformation strategy. Thus, in *C. reinhardtii* algae, the synthesis and metabolism of squalene, a high value-added molecule, are dependent on two enzymes, squalene synthase and squalene epoxidase. By blocking the gene for squalene epoxidase by genetic engineering, the accumulation of squalene (0.9–1.1 μg/mg cell dry mass) produced by squalene synthase is induced in the microalgae (Kajikawa *et al.* 2015).

Transgenesis is also used to increase the production of pigments, such as carotenoids by certain algae such as *Haematococcus pluvialis*, or the production of hormones from *C. reinhardtii* (Beachman *et al.* 2017) (see Table 6.4).

Species	Transgene encoding	Application	Expected benefit
Nannochloropsis salina	DGA1 (diglyceride acyltransferase)	Production of storage lipids (triacylglycerides)	Accumulation of lipids for production of biofuel
Chlamydomonas reinhardtii	Erythropoietin	Synthesis of hormone controlling the production of red blood cells	Synthesis of human hormone outside of animal models
Chlorella vulgaris	hGH	Production of human growth hormone	Synthesis of human hormone outside of animal models
Haematococcus pluvialis	pds	Production of phytoene desaturase	Enhanced synthesis of carotenoids
Phaeodactylum tricornutum	Elongase Desaturase	Biosynthesis of long-chain polyunsaturated fatty acids	Increased production of polyunsaturated fatty acids for human nutrition and aquaculture

Table 6.4. *Examples of transgenesis and their areas of application (Beachman* et al. *2017)*

One of the constraints associated with the use of transgenic microalgae remains, as with non-genetically modified algae, the cultivation conditions, and the avoidance of sources of contamination by other microorganisms.

Certain genetic modifications can respond to this problem of exogenous contamination of crops. Thus, in *C. reinhardtii* algae, the introduction of the ptxD gene coding for phosphite oxidoreductase, which oxidizes phosphite (HPO_3^-) to phosphate ($H_2PO_4^{2-}$), allows the transformed alga to grow on a medium whose sole source of phosphorus is phosphite. In this particular context, other potentially contaminating species lacking this gene will not grow, and therefore the culture will suffer little contamination (Loera-Quezada *et al.* 2016).

Transgenesis applied to microalgae has advantages and disadvantages compared to the production of other genetically modified organisms (GMOs).

The first advantage lies in the method of production. Indeed, microalgae can be cultivated in a closed system such as photobioreactors. The phenomena of intra- or interspecific gene flow with exogenous species from the external environment are theoretically null, which is far from being the case with terrestrial plants genetically modified and cultivated in an open environment.

Microalgae are microorganisms with a short generation time, which, depending on the conditions, can range from 3 to 11 hours for a model species such as *Chlamydomonas reinhardtii* (Sueoka 1960; Donnan *et al.* 1985). Such a yield in doubling the biomass is not possible with terrestrial plants whose cultivation requires several months.

One of the other advantages of transgenesis applied to microalgae lies in the total freedom from the use of animal sources to produce hormones or proteins of human or animal origin. This approach eliminates any risk of contamination by viruses or pathogens that could contaminate the original source.

Transgenesis applied to microalgae or cyanobacteria remains a promising avenue of development given the ease of cultivation of this resource, particularly in closed environments.

However, many locks remain to be lifted. The first is economic, as there is little data on the economic feasibility of such a biotechnological development, even from model species such as *C. reinhardtii*. The other lock concerns the regulatory aspects that could be applied. As such, there are notable differences of interpretation on the regulatory definition of a genetically modified organism between the United States and Europe.

In Europe, the terminology "genetically modified" is applied to any organism that has undergone an introduction, removal of genes or modification of specific parts of its genome by genetic engineering techniques. On the contrary, organisms genetically transformed as a result of random mutation techniques are not covered by European regulations on GMOs (Beacham *et al.* 2017). In the United States, the Department of Agriculture has not retained the status of GMO for a fungus genetically modified using the CRISPR/Cas 9. The argument is mainly based on the fact that the plants transformed by this molecular biology technique had not previously been listed as GMOs.

More generally, European regulations cover many aspects, such as the release of GMOs into the environment, the protection of the environment, the repair of damage caused by escapes and the introduction of GMOs in food or feed. The production of genetically modified algae is therefore subject to these regulations in Europe, which introduces an economic constraint to which other countries around the world are not subject.

Last but not least, public acceptance of GMOs remains the last, but not the least, hurdle. The opposition of one part of the public could have an impact on the development of a transgenic algae production chain. However, this observation deserves to be nuanced with regard to the oppositions expressed, which concern more the use of GMOs for food purposes and less for the production of therapeutic molecules.

Conclusion

Microalgae and cyanobacteria are the primary suppliers of organic matter and oxygen in our global ecosystem. In particular, they are the first link in the oceanic food chain. This fundamental role is recognized through the expression "primary producers", which is often attributed to them. Their use in animal feed, in certain sectors such as aquaculture, has a traditional character. This is particularly the case in France, where Marennes d'Oléron oysters are matured in tanks rich in diatoms belonging to the species *Haslea ostrearia*. Microalgae and especially the cyanobacterium *Spirulina* are also foods traditionally used in human nutrition. These resources have been mainly consumed by populations restricted to certain regions of Central America, Africa and Asia. The globalization of food practices has, however, allowed the development of the use as nutritional supplements of species belonging to the genera *Chlorella* and *Arthrospira* (Spirulina). This indicates that microalgae and cyanobacteria are foods well known to humankind. It gives them the status of food of the past and present. Concerning the future, this resource could also gain a status of food of the future. Their relative richness in proteins and polyunsaturated fatty acids make them prime candidates for the development of new sources of nutrients that are complementary or even alternative, under certain conditions, to those from animal resources. In terms of complementarity, this resource can provide certain populations with a supplement in terms of proteins and essential fatty acids.

This last aspect is particularly addressed in many projects supported by the FAO in Africa (Chad, Ethiopia), which works for the development of quality spirulina production chains. In Western countries, the emergence of new consumption practices (e.g. veganism and vegetarianism) or new dietary habits (e.g. reduced animal protein intake) opens up prospects for the use of microalgae in human food. Such an evolution would allow this resource to reach the new status of food of the future.

Due to their metabolism, microalgae and cyanobacteria are cellular factories producing very original molecules, such as phycobiliproteins that can be used in food coloring or immunofluorescent reagents. Their metabolic plasticity also makes it possible to synthesize, in a fairly well-controlled manner, molecules of interest useful for the production of biofuel. However, the production of biofuels from microalgae is currently limited by the economic feasibility of such an approach. This last parameter could nevertheless evolve favorably with regard to the progressive rarefaction of fossil fuels. In this context, algal biomass, due to its renewable nature, is a potential candidate to become, like other plant biomasses, a "green" source of fuels.

The cell factory concept, as previously described, will probably be modified in the future by the development of cell factories from genetically modified microalgae. Already validated by research, this new concept, which is well proven for the production of antibodies, antigens and hormones, will probably have to be extended to the production of new therapeutic molecules.

In the current quest for new vaccine proteins in response to epidemics and even pandemics, genetically modified microalgae are an interesting avenue for the development of new antiviral strategies.

References

Abdel-Tawwab, M. and Ahmad, M.H. (2009). Live Spirulina (*Arthrospira platensis*) as a growth and immunity promoter for Nile tilapia, *Oreochromis niloticus* (L.) challenged with pathogenic. *Aeromonas Hydrophila*, 40, 1037–1046.

Adam, F., Albert-Vian, M., Peltier, G., Chemat, F. (2012). "Solvent-free" ultrasound-assisted extraction of lipids from fresh microalgae cell: A green, clean and scalable process. *Bioresource Technology*, 114, 457–465.

Anderson, D.M., Glibert, P.M., Burkholder, J.M. (2002). Harmful algal blooms and eutrophication: Nutrient sources, composition, and consequences. *Estuaries*, 25, 704–726.

Andrich, G., Nesti, U., Venturi, F., Zinnai, A., Fiorentini, R. (2005). Supercrical fluid extraction of bioactive lipids from the microalga *Nannochloropsis* sp. *European Journal Lipid Sciences and Technology*, 107, 381–386.

Aritzia, E.V., Andersen, R.A., Sogin, M.L. (1991). A new phylogeny for chromophyte algae using 16S-like rRNA sequences from *Mallomonas papillosa* (Synurophyceae) and *Tribonema aequale* (Xanthophyceae). *Journal of Phycology*, 27, 428–436.

Batista, A.P., Gouveia, L., Bandarra, N.M., Franco, J.M., Raymundo, A. (2013). Comparison of microalgal biomass profiles as novel functional ingredient for food products. *Algal Research*, 2, 164–173.

Beacham, T.A., Sweet, J., Allen, M. (2017). Large scale cultivation of genetically modified microalgae: A new era for environmental risk. *Algal Research*, 25, 90–100.

Becerra-Celis, G. (2009). Proposition de stratégies de commande pour la culture de microalgue dans un photobioréacteur continu. Thesis, Centrale Paris, SupElec, Paris.

Becker, E.W. (2013a). Microalgae for aquaculture: Nutritional aspect. In *Handbook of Microalgal Culture: Applied Phycology and Biotechnology*, Richmond, A. and Hu, Q. (eds). John Wiley & Sons/Blackwell Publishing, Hoboken.

Becker, E.W. (2013b). Microalgae for human and animal nutrition. In *Handbook of Microalgal Culture: Applied Phycology and Biotechnology*, Richmond, A. and Hu, Q. (eds). John Wiley & Sons/Blackwell Publishing, Hoboken.

Bhattacharya, D., Medlin, L., Wainright, P.O., Ariztia, E.V., Bideau, C., Stickel, S.K., Sogin, M.L. (1992). Algae containing chorophylls a+c are paraphyletic molecular evolutionary analysis of the Chromophyta. *Evolution*, 46, 1801–1817.

Borowitzka, M.A. (2005). Culturing microalgae in outdoor ponds. In *Algal Culturing Techniques*, Andersen, R.A. (ed.). Academic Press/Elsevier, London.

Borowitzka, M.A. (2018a). Biology of microalgae. In *Microalgae in Health and Disease Prevention*, Levine, I. and Fleurence, J. (eds). Academic Press/Elsevier, London.

Borowitzka, M.A. (2018b). Microalgae in medicine and human health: A historical perspective. In *Microalgae in Health and Disease Prevention*, Levine, I. and Fleurence, J. (eds). Academic Press/Elsevier, London.

Bourre, J.M., Oalan, O., Trygve Berg, L. (2006). Les teneurs en acides gras oméga-3 des saumons Atlantique sauvages (d'Écosse, Irlande et Norvège) comme références pour ceux d'élevage. *Médecine et Nutrition*, 42, 36–49.

Brennan, L. and Owende, P. (2010). Biofuels from microalgae – A review of technologies for production, processing, and extractions of biofuels and co-products. *Renewable and Suistainable Energy Reviews*, 14, 557–577.

Brin, A.J. and Goutelard, N. (1994). Composition cosmétique ou pharmaceutique anti-radicaux libres pour application topique. Demande de brevet européen, 0629397, A1.

Brüll, L.P., Huang, Z., Thomas-Oates, J.E., Paulsen, B.S., Cohen, E.H., Michaelsen, T.E. (2000). Studies of polysaccharides from three edible species of *Nostoc* (Cyanobacteria) with different colony morphologies: Structural characterization and effect on the complement system of polysaccharides from *Nosto*c commune. *Journal of Phycology*, 36, 871–881.

Bruneell, C., Lemahieu, C., Fraeye, I., Ryckebosch, E., Muylaert, K., Buyse, J., Foubert, I. (2013). Impact of microalgal feed supplementation on omega-3 fatty acid enrichment of hen eggs. *Journal of Functional Foods*, 5, 897–904.

Cadoret, J.P., Bardor, M., Lerouge, P., Cabigliera, M., Henriquez, V., Carlier, A. (2008). Les microalgues : usines cellulaires productrices de molécules commerciales recombinantes. *MS – médecine sciences*, 24, 375–384.

Carey, C.C., Ibelings, B.W., Hoffman, E.P., Hamilton, D.P., Brookes, J.D. (2012). Eco-physiological adaptations that favour freshwater cyanobacteria in a changing climate. *Water Research*, 46, 1394–1407.

Cha, K.H., Koo, S.Y., Lee, D. (2008). Antiproliferative effects of carotenoids extracted from *Chlorella ellipsoidea* and *Chlorella vulgaris* on human colon cancer cells. *Journal of Agricultural and Food Chemistry*, 56, 10521–10526.

Cha, K.H., Lee, H.J., Koo, S.Y., Song, D.G., Lee, D.U., Pan, C.H. (2010). Optimization of pressurized liquid extraction of carotenoids and chlorophylls from *Chlorella vulgaris*. *Journal of Agricultural and Food Chemistry*, 58, 793–797.

Charlson, R.J., Lovelock, J.E., Andreae, M.O., Warren, S.G. (1987). Oceanic phytoplankton, atmospheric Sulphur, cloud albedo and climate. *Nature*, 326, 655–661.

Chisti, Y. (2018). Society and microalgae: Understanding the past and present. In *Microalgae in Health and Disease Prevention*, Levine, I. and Fleurence, J. (eds). Academic Press/Elsevier, London.

Cohen, Z. and Khozin-Goldberg, I. (2010). Searching for polyunsatured fatty acid-rich photosynthetic microalgae. In *Single Cell Oils: Microbial and Algal Oils*, Cohen, Z. and Ratledge, C. (eds). AOCS Press, Urbana.

Coutteau, P. and Sorgeloos, P. (1992). The use of algal substitutes and the requirement for live algae in the hatchery and nursery rearing of bivalves molluscs: An international survey. *Journal of Shellfish Research*, 11, 467–471.

Davis, E.W. (1983). The ethnobiology of the Haitian zombi. *Journal of Ethnopharmacology*, 9, 85–104.

Davis, R., Aden, A., Pienkos, P. (2011). Techno-economic analysis of autotrophic microalgae for fuel production. *Applied Energy*, 88, 3524–3531.

De Godos, I., Mendoza, J.L., Acien, F.G., Molina, E., Banks, C.J., Heaven, S., Rogalla, F. (2014). Evaluation of carbon dioxide mass transfer in raceway reactors for microalgae culture using flue gases. *Bioresource Technology*, 153, 307–314.

De Vargas, C., Audic, S., Henry, N., Decelle, J., Mahé, F., Logares, R., Lara, E., Berney, C., Le Bescot, N., Probert, I. *et al.* (2012). Eukaryotic plankton diversity in the sunlit ocean. *Science*, 348.

Deenu, A., Naruenartwongsakul, S., Kim, S.M. (2013). Optimization and economic evaluation of ultrasound extraction of lutein from *Chlorella vulgaris*. *Biotechnology and Bioprocess Engineering*, 18, 1151–1162.

Delpeuch, F., Joseph, A., Cavelie, C. (1976). Consommation alimentaire et apport nutritionnel des algues bleues (*Oscillatoria platensis*) chez quelques populations du Kanem (Tchad). *Annales nutritionnelles alimentaires*, 29, 497–516.

Demuez, M., Mahdy, A., Tomas-Pejo, E., Gonzales-Fernadez, C., Ballesteros, M. (2015). Enzymatic cell disruption of microalgae biomass in biorefinery processes. *Biotechnology and Bioengineering*, 112, 1955–1966.

Dewi, I.C., Falaise, C., Hellio, C., Bourgougnon, N., Mouget, J.L. (2018). Anticancer, antiviral, antibacterial and antifungal properties in microalgae. In *Microalgae in Health and Disease Prevention*, Levine, I. and Fleurence, J. (eds). Academic Press/Elsevier, London.

Dey, S. and Rathod, V.K. (2013). Ultrasound assisted extraction of β-carotene from *Spirulina platensis*. *Ultrasonics Sonochemistry*, 20, 271–276.

DGCCRF (2019). ALGUES. Liste des algues pouvant être employées dans les compléments alimentaires. Document, Nutrition et information des consommateurs. Secteur “Compléments alimentaires”, 1–5.

Dodge, J.D. (1985). *Atlas of Dinoflagellates. A Scanning Electronic Microscope Survey*. Farrand Press Editions, London.

Donnan, L., Carvill, E.P., Gilliland, T.J., John, P.C.I. (1985). The cell cycles of *Chlamydomonas* and *Chlorella*. *New Phytologist*, 99, 1–40.

Duarte-Santos, T., Mendoza-Martin, J.L., Acien Fernandez, F.G., Molina, E., Vieira-Costa, J.A., Heaven, S. (2016). Optimization of carbon dioxide supply in raceway reactors: Influence of carbon dioxide molar fraction and gas flow rate. *Bioresource Technology*, 212, 72–81.

Dufossé, L., Galaup, P., Yaron, A., Arad, S.M., Blanc, P., Murthy, K.N.C., Ravishankar, G. (2005). Microorganisms and microalgae as source of pigments for food use: A scientific oddity or an industrial reality? *Trends in Food Sciences and Technology*, 16, 389–406.

Dumay, J., Morançais, M., Munier, M., Le Guillard, C., Fleurence, J. (2014). Phycoerythrins: Valuable proteinic pigments in red seaweeds. In *Advances in Botanical Research*, Bourgougnon, N. (ed.). Academic Press/Elsevier, London.

Dunahay, T.G., Jarvis, E.E, Roessler, P.G. (1995). Genetic transformation of the diatoms *Cyclotella cryptica* and *Navicula saprophila*. *Journal of Phycology*, 31, 1004–1012.

Dupré, C. and Grizeau, D. (1994). Procédé de production de microalgues. Brevet international, WO 94/07993.

Durmaz, Y., Monteiro, M., Bandarra, N., Gökpinar, S., Isik, O. (2007). The effect of low temperature on fatty acid composition and tocopherols of the red microalga *Porphyridium cruentum*. *Journal of Applied Phycology*, 19, 223–227.

European Parliament (1997). Réglement (CE) n°258/97 du Parlement européen et du Conseil du 27 janvier 1997 relatif aux nouveaux aliments et aux nouveaux ingredients alimentaires. *Journal officiel des Communautés européennes*, n°L 43, 1–6.

European Union (2002). Directive 2002/46/CE du Parlement européen et du Conseil du 10 juin 2002 relative aux rapprochements des législations des États membres concernant les compléments alimentaires (texte présentant l'intérêt pour l'EEE). *Journal officiel des Communautés européennes*, L 183, 51–57.

FAO (2012). Fisheries and aquaculture department, Statistic and information service. FishStatJ: Universal software for fishery statistical time series.

FAO (2018a). *La situation mondiale des pêches et de l'aquaculture 2018 : atteindre les objectifs de développement durable*. FAO, Rome.

FAO (2018b). OECD-FAO Agricultural Outlook 2018–2027. Report, FAO, Rome.

Faust, M.A. and Gulledge, R.A. (2002). Identifying harmful marine dinoflagellate. *Contributions from the United States National Herbarium*, 42, 1–144.

Fernandez, J.J., Candenas, M.L., Souto, M.L., Trujillo, M.M., Norte, M. (2002). Okadaic acid, useful tool for studying cellular processes. *Current Medicinal Chemistry*, 9, 229–262.

Fleurence, J. (2016). Seaweed as food. In *Seaweed in Health and Disease Prevention*, Fleurence, J. and Levine, I. (eds). Academic Press/Elsevier, London.

Fleurence, J. (2018). *Les algues alimentaires : bilan et perspectives*. Lavoisier, Paris.

Fleurence, J. and Joubert, Y. (2006). Procédé d'extraction d'ADN incluant une étape de digestion enzymatique. Patent, national registration number 0407278.

Fleurence, J. and Levine, I. (2018). Antiallergic and allergic properties. In *Microalgae in Health and Disease prevention*, Levine, I. and Fleurence, J. (eds). Academic Press/Elsevier, London.

Fujitani, N., Sakaki, S., Yamaguchi, Y., Takenaka, H. (2001). Inhibitory effects of microalgae on the activation of hyaluronidase. *Journal of Applied Phycology*, 13, 489–492.

Gallardo-Rodriguez, J., Sanchez-Miron, A., Garcia-Camacho, F., Lopez-Rosales, L., Chisti, Y., Molina-Grima, E. (2012). Bioactives from microalgal dinoflagellates. *Biotechnology Advances*, 30, 1673–1684.

Garcia Bueno, N. (2015). Valorisation de la macroalgue proliférante *Grateloupia turuturu* dans l'élevage de l'ormeau européen *Haliotis tuberculata*. University thesis, Nantes.

Geresh, S., Adin, I., Yarmolinsky, E., Karpasas, M. (2002). Characterization of the extracellular polysaccharide of *Porphyridium* sp.: Molecular weight determination and rheological properties. *Carbohydrate Polymers*, 50, 183–189.

Gerken, H.G., Donohe, B., Knoshaug, E.P. (2013). Enzymatic cell wall degradation of *Chlorella vulgaris* and other microalgae for biofuels production. *Planta*, 237, 239–253.

Ghasemi, Y., Rasoul-Amini, S., Naseri, A.T., Montazeri-Najafabady, N., Mobasher, M.A., Dabbagh, F. (2012). Microalgae biofuel potentials (Review). *Applied Biochemistry and Microbiology*, 48, 126–144.

Gouveia, L., Gomes, E., Empis, J. (1996). Potential use of microalga (*Chlorella vulgaris*) in the pigmentation of rainbow trout (*Oncorhyncus mykiss*) muscle. *Zeitung Lebensm Unters Forsch*, 202, 75–79.

Gouveia, L., Raymundo, A., Batista, A.P., Sousa, I., Empis, J. (2006). *Chlorella vulgaris* and *Haematococcus pluvialis* biomass as colouring and antioxidant in food emulsion. *European Food Research Technology*, 222, 362–367.

Gouveia, L., Batista, A.P., Miranda, A., Empis, J., Raymundo, A. (2007). *Chlorella vulgaris* biomass used as colouring source in traditional butter cookies. *Innovative Food Science and Emerging Technologies*, 8, 433–436.

Greenhalgh, M. and Ovenden, D. (2009). *Guide de la vie des eaux douces. Les plantes, les animaux et les empreintes. Les guides du naturaliste*. Delachaux et Niestlé Éditions, Paris.

Gudin, C. (2013). *Histoire naturelle des microalgues*. Odile Jacob, Paris.

Gudin, C. and Chaumont, D. (1991). Cell fragility – The key problem of microalgae mass production in closed photobioreactors. *Bioresource Technology*, 38, 145–151.

Gudin, C. and Trezzy, C. (1994). Procédé de production et d'extraction de superoxyde-dismutases thermostables à partir d'une culture de microorganismes photosynthétiques. European patent application, 0628629, A1.

Guiry, M.D. (2012). How many species of algae are there? *Journal of Phycology*, 48, 1057–1063.

Haché, R., Dumas, A., Thumbi, D., Forward, B.S., Mallet, M. (2017). Effect of live algae used as green water on survival, growth, behavior, ontogeny and bacterial profile of lobster larvae (*Homarus americanus* Milne Edwards). *Aquaculture Research*, 48, 581–593.

Hagen, N., Cantin, L., Constant, J., Haller, T., Blaise, G., Ong-Lam, M., du Souich, P., Korz, W., Lapointe, B. (2017). Tetrodotoxin for moderate to severe cancer-related pain: A multicentre, randomized, double-blind, placebo-controlled, parallel-design trial. *Pain Research and Management*, ID 7212713.

Han, D., Deng, Z., Lu, F., Hu, Z. (2013). Biology and biotechnology of edible *Nostoc*. In *Handbook of Microalgal Culture: Applied Phycology and Biotechnology*, Richmond, A. and Hu, Q. (eds). John Wiley & Sons/Blackwell Publishing, Hoboken.

Harder, R. and von Witsch, R. (1942). Über die Massenkultur von Diatomeen. *Berichte der Deutschen Botanischen Gesselschaft*, 60, 146–152.

Hase, R., Oikawa, H., Sasao, C., Morita, M., Watanabe, Y. (2000). Photosynthetic production of microalgal biomass in the raceway system under greenhouse conditions in Sendai city. *Journal of Biosciences and Bioengineering*, 89(2), 157–163.

Hayashi, T., Hayashi, K., Maeda, M., Kojima, I. (1996). Calcium Spirulan, an inhibitor of enveloped virus replication, from a blue-green alga *Spirulina platensis*. *Journal of Natural Products*, 59, 83–87.

Helm, M.M. and Bourne, N. (2006). Écloserie de bivalves. Un manuel pratique. In *FAO Document technique sur les pêches*, Lovatelli, A. (ed.). FAO, Rome.

Huang, Z., Liu, Y., Paulsen, B.S., Klaveness, D. (1998). Studies of polysaccharides from three edible *Nostoc* (Cyanobacteria) with different colony morphologies: Comparison of monosaccharide composition and viscosities of polysaccharides from field colonies. *Journal of Phycology*, 34, 962–968.

Iltis, A (1980). Les algues. In *Flore et faune aquatiques de l'Afrique Sahelo-soudanienne*, Durand, J.R. and Lévêque, C. (eds). IRD Éditions, Paris.

Iman, S.H., Buchanan, M.J., Shin, H.C., Snell, W.J. (1985). The Chlamydomonas cell wall: Characterization of the wall framework. *The Journal of Cell Biology*, 101, 1599–1607.

Islam, M.M., Begum, S., Lin, L., Okamura, A., Du, M., Fujimara, S. (2002). Synergistic cytotoxic effect between serine-threonine phosphatase inhibitors and 5-fluorouracil: A novel concept for modulation of cytotoxic effect. *Cancer Chemotherapy and Pharmacology*, 49, 111–118.

Jarisoa, T. (2005). Adaptation de la spiruline du sud de Madagascar à la culture en eau de mer. Mise au point des structures de production à l'échelle villageoise. University thesis, Tolaria.

Joannes, C., Sipaut, C.S., Dayou, J., Yasir, S.M., Mansa, R.F. (2015). Review paper on cell membrane electroporation of microalgae using electric field treatment method for microalgae lipid extraction. *Materials Science and Engineering*, 78, 1–8.

Jubeau, S., Marchal, L., Pruvost, J., Jaouen, P., Legrand, J., Fleurence, J. (2013). High pressure disruption: A two-step treatment for selective extraction of intracellular components from the microalga *Porphyridium cruentum*. *Journal of Applied Phycology*, 25, 983–989.

Kajikawa, M., Kinohira, S., Ando, A., Shimoyama, M., Kato, M., Fukuzawa, H. (2015). Accumulation of Squalene in a microalga *Chlamydomonas reinhardtii* by genetic modification of squalene synthase and squalene epoxidase genes. *PLOS One*, 1–21.

Keris-Sen, U.D., Sen, U., Soydemir, G., Gurol, M.D. (2014). An investigation of ultrasound effect of microalgal cell integrity and lipid extraction efficiency. *Bioresource Technology*, 152, 407–413.

Kim, H.M. (2000). Anti-allergic drugs from Oriental medicine. *International Journal of Medicine*, 1, 1–7.

Kim, Y.S., Ahn, K.H., Kim, S.Y., Jeong, J.W. (2009). Okadaic acid promotes angiogenesis via activation of hypoxia-inductible factor 1. *Cancer Letters*, 276, 102–108.

Kitada, K., Machmudah, S., Sasaki, M., Goto, M., Nakashima, Y., Kumamoto, S., Hasegawa, T. (2009). *Journal of Chemical Technology and Biotechnology*, 84, 657–661.

Kodama, M., Sato, S., Sakamoto, S., Ogata, T. (1996). Occurrence of Tetrodotoxin in *Alexandrium tamarense*, a causative dinoflagellate of paralytic shellfish poisoning. *Toxicon: Official Journal of the International Society on Toxinology*, 10, 1101–1105.

Kumar, V., Muthuraj, M., Palabhanvi, B., Ghoshal, A.K., Das, D. (2014). High cell density lipid rich cultivation of a novel microalgal isolate *Chlorella sorokiniana* FC6 IITG in a single-stage fed-batch mode under mixotrophic condition. *Bioresource Technology*, 170, 115–124.

Kurmaly, K., Jones, D.A., Yule, A.B., East, J. (1989). Comparative analysis of the growth and survival of *Penaeus monodon* (Fabricius) larvae, from protozoa 1 to postlarva 1 on live feeds, artificial diets and on combinations of both. *Aquaculture*, 81, 27–45.

Lecointre, G. and Le Guyader, H. (2016). *Classification phylogénétique du vivant*, volume 1, 4th edition. Belin, Paris.

Lee, S.Y., Show, P.L., Ling, T.C., Chang, J.S. (2017). Single-step disruption and protein recovery from *Chlorella vulgaris* using ultrasonication and ionic liquid buffer aqueous solutions as extractive solvents. *Biochemical Engineering Journal*, 124, 26–35.

Légifrance (2006). Décret n°2006-352 du 20 mars 2006 relatif aux compléments alimentaires. NOR ECOC0500166D [Online]. Available at: https:/www. legifrance. gouv.fr.

Li, C.W. and Volcani, B. (1987). Four new apochlorotic diatoms. *British Phycology Journal*, 22, 375–382.

Loera-Quezada, M.M, Leyva-Gonzalez, M.A, Velazquez-Juarez, G., Sanchez-Calderon, L., Do Nascimento, M., Lopez-Arredonda, D., Herrera-Estrella, L. (2016) A novel genetic engineering platform for the effective management of biological contaminants for the production of microalgae. *Plant Biotechnology Journal*, 14, 2066–2076.

Loir, M. (2004). *Guide des diatomées. Les guides du naturaliste*. Delachaux et Niestlé Éditions, Paris.

Lovelock, J.E., Maggs, R.J., Rasmussen, R.A. (1972). Atmospheric dimethyl sulphide and the natural Sulphur cycle. *Nature*, 237, 4.

Lu, Y.M., Xiang, W.Z., We, Y.H. (2011). *Spirulina* (*Arthrospira*) industry in Inner Mongolia of China: Current status and prospects. *Journal of Applied Phycology*, 23, 265–269.

Lucas, P.L., Mathieu-Rivet, E., Song, P.C.T., Oltmanns, A., Loutelier-Bourhiss, C., Plasson, C., Alfonso, C., Hippler, M., Lerouge, P., Mati-Baouche, N., Bardor, M. (2020). Multiple xylotransferases heterogeneously xyloxylate protein N-linked glycans in *Chlamydomonas reinhardtii. The Plant Journal*, 102, 230–245.

Mabeau, S. and Fleurence, J. (1993). Seaweed in food products: Biochemical and nutritional aspects. *Trends in Food Science and Technology*, 4, 103–107.

Maeda, Y., Yoshino, T., Matsunaga, T., Matsumoto, M., Tanaka, T. (2018). Marine microalgae for production of biofuels and chemicals. *Current Opinion in Biotechnology*, 50, 111–120.

Malanga, G. and Puntarulo, S. (1995). Oxidative stress and antioxidant content in Chlorella vulgaris after exposure to ultraviolet-B radiation. *Physiologia plantarum*, 94, 672–679.

Mann, D.G. and Vanormelingen, P. (2013). An inordinate fondness? The number, distribution and origins of diatoms species. *Journal of Eukaryotic Microbiology*, 60, 414–420.

Marcati, A., Ursu, A.V., Laroche, C., Soanen, N., Marchal, L., Jubeau, S., Djelveh, G., Michaud, P. (2014). Extraction and fractionation of polysaccharides and B-phycoerythrin from the microalga *Porphyridium cruentum* by membrane technology. *Algal Research*, 5, 258–263.

Mata, T.M., Martins, A.A., Caetano, N.S. (2010). Microalgae for biodiesel production and other applications: A review. *Renewable and Sustainable Energy Reviews*, 14, 217–232.

Mathieu-Rivet, E., Scholz, M., Arias, C., Dardelle, F., Schulze, S., Le Mauff, F., Teo, G., Hochmal, A.K., Blanco-Rivero, A., Loutelier-Bourhiss, C., Kiefer-Meyer, M.C., Fufezan, C., Burel, C., Lerouge, P., Martinez, F., Bardor, M., Hippler, M. (2013). Exploring the N-glycosylation patway in *Chlamydomonas reinhardtii* unravels novel complex structures. *Molecular & Cellular Proteomics*, 12, 3160–3183.

Mayfield, P.S., Franklin, S.E., Lerner, R.A. (2003). Expression and assembly of a fully active antibody in algae. *PNAS*, 100, 438–442.

Medcalf, D.G., Scott, J.R., Brannon, J.H., Hemerick, G.A. (1975). Some structural features and viscometric properties of the extracellular polysaccharides from *Porphyridium cruentum*. *Carbohydrate Research*, 44, 87–96.

Melis, A. and Happe, T. (2001). Hydrogen production. Green algae as a source of energy. *Plant Physiology*, 127, 740–748.

Mendes, R.L., Fernandes, H.L., Coelho, J.P., Reis, E.C., Cabral, J.M.S., Novais, J.M., Palavra, A.F. (1995). Supercritical CO_2 extraction of carotenoids and lipids from *Chlorella vulgaris*. *Food Chemistry*, 53, 99–103.

Menéndez, J.M.B., Arenillas, A., Menéndez Diaz, J.A., Boffa, L., Mantegna, S., Binello, A., Cravatoo, G. (2014). Optimization of microalgae oil extraction under ultrasound and microwave irradiation. *Journal of Chemical Technology and Biotechnology*, 89, 1779–1784.

Mimouni, V., Couzinet-Mossion, A., Ulmann, L., Wielgosz-Collin, G. (2018). Lipids from microalgae. In *Microalgae in Health and Disease Prevention*, Levine, I. and Fleurence, J. (eds). Academic Press/Elsevier, London.

Minkova, K.M., Toshkova, R.A., Gardeva, E.G., Tchorbajieva, M.I., Ivanora, N.J., Yossifova, L.S., Gigova, L.G. (2011). Antitumor activity of B-phycoerythrin from *Porphyridium cruentum*. *Journal of Pharmacy Research*, 4, 1480–1482.

Misra, H.P. and Fridovich, I. (1977). Purification and properties of superoxide dismutase from a red alga, *Porphyridium cruentum. The Journal of Biological Chemistry*, 18, 6421–6423.

Molina, M., Fernandez, J., Acien, F.G., Chisti, Y. (2001). Tubular photobioreactor design for algal cultures. *Journal of Biotechnology*, 92, 113–131.

Montalescot, V., Rinaldi, T., Touchard, R., Jubeau, S., Frappart, M., Jaouen, P., Bourseau, P., Marchal, L. (2015). Optimization of bead milling parameters for cell disruption of microalgae: Process modelling and application to *Porphyridium cruentum* and *Nannochloropsis oculata*. *Bioresource Technology*, 196, 339–346.

Morançais, M., Mouget, J.L., Dumay, J. (2018). Proteins and pigments. In *Microalgae in Health and Disease Prevention*, Levine, I. and Fleurence, J. (eds). Academic Press/Elsevier, London.

Moreau, D., Tomasoni, C., Jacquot, C., Kaas, R., Le Guedes, R., Cadoret, J.P., Muller-Feuga, A., Kontiza, I., Viagas, C., Roussis, V., Roussakis, C. (2006). Cultivated microalgae and carotenoid fucoxanthin from *Odontella aurita* as potent anti-proliferative agents in bronchopulmonary and epithelial cell lines. *Environmental Toxicology and Pharmacology*, 22, 97–103.

Moreno-Garcia, L., Adjallé, K., Barnabé, S., Raghavan, G.S.V. (2017). Microalgae biomass production for a refenery system: Recent advances and the way towards sustainability. *Renewable and Sustainable Energy Reviews*, 76, 493–506.

Morimoto, T., Nagatsu, A., Murakami, N., Sakakibara, J., Tokuda, H., Nishino, H., Iwashima, A. (1995). Anti-tumor promoting glyceroglycolipids from the green alga, *Chlorella vulgaris*. *Phytochemistry*, 40, 1433–1437.

Morito, T., MacGregor, J.T., Hayashi, M. (2011). Micronucleus assays in rodent tissues other than bone marrow. *Mutagenesis*, 26, 223–230.

Muller-Feuga, A. (2000). The role of microalgae in aquaculture: Situation and trends. *Journal of Applied Phycology*, 12, 527–534.

Muller-Feuga, A. (2013). Microalgae for aquaculture: The current global situation and future trends. In *Handbook of Microalgal Culture: Applied Phycology and Biotechnology*, Richmond, A. and Hu, Q. (eds). John Wiley & Sons/Blackwell Publishing, Hoboken.

Muller-Feuga, A., Le Guédes, R., Hervé, A., Durand, P. (1998). Comparison of artificial light photobioreactors and other production system. *Journal of Applied Phycology*, 10, 83–90.

Murthy, K.N.C., Vanitha, A., Rajesha, J., Swamy, M.M., Sowmya, P.R., Ravishankar, A.G. (2005). *In vivo* antioxidant activity of carotenoids from *Dunaliella salina* – A green microalga. *Life Sciences*, 76, 1381–1390.

Najdenski, H.M., Gigova, L.G., Illiev, I.I., Pilarski, P.S., Lukavsky, J., Tsvetkova, I.V., Ninova, M.S., Kussovski, V.K. (2013). Antibacterial and antifungal activities of selected microalgae and cyanobacteria. *International Journal of Food Science and Technology*, 48, 1533–1540.

Nakanishi, A., Aikawa, S., Ho, S.H., Chen, C.Y., Chang, J.S., Hasunuma, T., Kondo, A. (2014). Development of lipid productivities under different CO_2 conditions of marine microalgae *Chlamydomonas* sp. JSC4. *Bioresource Technology*, 152, 247–252.

Nazih, H. and Bard, J.M. (2018). Microalgae in the human health: Interest as functional food. In *Microalgae in Health and Disease Prevention*, Levine, I. and Fleurence, J. (eds). Academic Press/Elsevier, London.

Nef, C. (2019). Métabolisme et interactions bactériennes en lien avec la vitamine B12 chez la microalgue haptophyte *Tisochrysis lutea*. University thesis, Nantes.

Nobre, B.P., Villalobos, F., Barragan, B.E., Oliveira, A.C., Batista, A.P., Marques, P.A.S.S., Mendes, R.L., Sonova, H., Palavra, A.F., Gouveia, L. (2013). A biorefinery from *Nannochloropsis* sp. microalga – extraction of oils and pigments. Production of biohydrogen from the leftover biomass. *Bioresource Technology*, 135, 128–136.

Pal, D., Khozin-Goldberg, I., Cohen, Z., Boussiba, S. (2011). The effect of light, salinity, and nitrogen availability on lipid production by *Nannochloropsis* sp. *Applied Microbiology and Biotechnology*, 90, 1429–1441.

Pasquet, V., Ulmann, L., Mimouni, V., Guihéneuf, F., Jacquette, B., Morant-Manceau, A., Tremblin, G. (2014). Fatty acids profile and temperature in the cultured marine diatom *Odontella aurita*. *Journal of Applied of Phycology*, 26, 2265–2271.

Peiretti, P.G. and Meineri, G. (2008). Effects of diets with increasing levels of *Spirulina platensis* on the performance and apparent digestibility in growing rabbits. *Livestock Science*, 118, 173–177.

Perez, R. (1997). *Ces algues qui nous entourent. Conception actuelle, rôle dans la biosphère, utilsations, culture*. IFREMER Éditions, Plouzané.

Philippis, R.D. and Vincenzini, M. (1998). Exocellular polysaccharides from cyanobacteria and their possible applications. *FEMS Microbiology Reviews*, 22, 151–175.

Phong, W.N., Show, P.L., Ling, T.C., Juan, J.C., Ng, E.P., Chang, J.S. (2018). Mild cell disruption methods for bio-functional proteins recovery from microalgae – Recent developments and future perspectives. *Algal Research*, 31, 506–516.

Ponis, E., Parisi, G., Robert, R. (2002). Valeur alimentaire de *Tetraselmis striata* et *T.chui* pour les larves de *Crassostrea gigas*. *Haliotis*, 31, 57–62.

Postma, P.R., Miron, T.L., Olivieri, G., Barbosa, M.J., Wijffels, R.H., Eppink, M.H.M. (2015). Mild disintegration of the green microalgae *Chlorella vulgaris* using bead milling. *Bioresource Technology*, 184, 297–304.

Pouvreau, J.B., Morançais, M., Taran, F., Rosa, P., Dufossé, L., Guérard, F., Pin, S., Fleurence, J., Pondaven, P. (2008). Antioxidant and free radical scavenging properties of Marennine, a blue-green polyphenolic pigment from the diatom *Haslea ostrearia* (Gaillon/Bory) Simonsen responsible for the naural greening of cultured oysters. *Journal of Agricultural and Food Chemistry*, 56, 6278–6286.

Radokovits, R., Jinkerson, R.E., Darzins, A., Posewitz, M. (2010). Genetic engineering of algae for enhanced biofuel production. *Eucaryotic Cell*, 9, 486–501.

Raposo, M.F.J., de Morais, A.M.M.B., de Morais, R.M.S.C. (2014). Influence of sulphate on the composition and antibacterial and antiviral properties of the exopolysaccharide from *Porphyridium cruentum. Life Sciences*, 101, 56–63.

Richmond, A., Lichtenberg, E., Stahl, B., Vonshak, A. (1990). Quantitative assessment of the major limitations on productivity of *Spirulina platensis* in open raceways. *Journal of Applied Phycology*, 2, 195–206.

Robert, R. and Trintignac, P. (1997). Microalgues et nutrition larvaire en écloserie de mollusques. *Haliotis*, 26, 1–13.

Rodrigues, M.A. and da Silva Bon, E.P. (2011). Evaluation of *Chlorella* (Chlorophyta) as source of fermentable sugars via cell wall enzymatic hydrolysis. *Enzyme Research*, 1–5.

Roney, B.R., Renhui, L., Banack, S.A., Murch, S., Honegger, R.M., Cox, P.A. (2009). Consumption of *fa cai Nostoc* soup: A potential for BMAA exposure from *Nostoc* cyanobacteria in China? *Amyotrophic Lateral Sclerosis*, 10, 44–49.

Ross, E. and Domini, W. (1990). The nutritional value of dehydrated, blue-green algae (*Spirulina platensis*) for poultry. *Journal Series, the Hawaii Institute of Tropical Agriculture*, 3338, 794–800.

Round, F.E., Crawford, R.M., Mann, D.J. (1990). *The Diatoms. Biology & Morphology of the Genera*. Cambridge University Press, Cambridge.

Safi, C., Zebib, B., Merah, O., Pontalier, P.Y., Vaca-Garcia, C. (2014). Morphology, composition, production, processing and applications of *Chlorella vulgaris*: A review. *Renewable and Sustainable Energy Reviews*, 35, 265–278.

Sanchez, M., Bernal-Castillo, J., Rozo, C., Rodriguez, I. (2003). *Spirulina* (*Arthrospira*) An edible microorganism: A review. *Universitas Scientiarum*, 8, 7–24.

Sannasimuthu, A., Kumaresan, V., Pasupuleti, M., Paray, B.A., Al-Sadoon, M.K., Arockiaraj, J. (2018). Radical scavenging property of a novel peptide derived from C-terminal SOD domain of superoxide dismutase enzyme in *Arthrospira platensis*. *Algal Research*, 35, 519–529.

Sardet, C. (2013). *Plancton. Aux origines du vivant*. Ulmer Éditions, Paris.

Scholz, M.J., Weiss, T.L., Jinkerson, R.E., Jing, J., Roth, R., Goodenough, U., Posewitz, M.C., Gerken, H.G. (2014). Ultrastructure and composition of the Nannochloropsis gadinata cell wall. *Eucaryotic Cell*, 13, 1450–1464.

Selosse, M.A. (2006). Animal ou végétal? Une distinction obsolète. *Pour la Science*, 66–72.

Sexton, J.P. and Lomas, M.W. (2018). Microalgal Systematics. In *Microalgae in Health and Disease Prevention*, Levine, I. and Fleurence, J. (eds). Academic Press/Elsevier, London, 71–107.

Sharma, O.P. (1986). Xanthophyceae (yellow-green algae). In *Texbook of Algae*, Sharma, O.P. (ed.). Tata McGraw Hill Publishing Company, New Delhi, 270–277.

Sharma, N.K. and Rai, A.K. (2011). Algal particles in the atmosphere. In *Encyclopedia of Environmental Health*, Nriagu, J.O. (ed.). Elsevier, London.

Sharma, N.K., Singh, S., Rai, A.K. (2006). Diversity and seasonal variation of algal particles in the atmosphere of subtropical city in India. *Environmental Research*, 102, 252–259.

Sierra, E., Acien, F.G., Fernandez, J.M., Garcia, J.L., Gonzalez, C., Molina, E. (2008). Characterization of a flat plate photobioreactor for the production of microalgae. *Chemical Engineering Journal*, 138, 136–147.

Sierra, L.S., Dixon, C.K., Wilken, L.R. (2017). Enzymatic cell disruption of the microalgae *Chlamydomonsa reinhardtii* for lipid and protein extraction. *Algal Research*, 25, 149–159.

Singh, J. and Gu, S. (2010). Commercialization potential of microalgae for biofuel production. *Renewable and Sustainable Energy Reviews*, 14, 2596–2610.

Singh, P., Guldhe, A., Kumari, S., Rawat, I., Bux, F. (2015). Investigation of combined effect of nitrogen, phosphorus and iron on lipid productivity of microalgae *Ankistrodesmus falcatus* KJ671624 using response surface methodology. *Biochemical Engineering Journal*, 94, 22–29.

Solovchenko, A.E., Khozin-Goldberg, I., Didi-Cohen, S., Cohen, Z., Merzlyak, M.N. (2008). Effects of light intensity and nitrogen starvation on growth, total fatty acids and arachidonic acid in the green microalga *Parietochloris incisa*. *Journal of Applied Phycology*, 20, 245–251.

Sournia, A., Chrétiennot-Dinet, M.J., Ricard, M. (1991). Marine phytoplankton: How many species in the world ocean? *Journal of Plankton Research*, 13, 1093–1099.

Spector, D.L. (1984). Dinoflagellates: An introduction. In *Dinoflagellates*, Spector, D.L. (ed). Academic Press, New York.

Spiden, E.M., Yap, B.H.J., Hill, D.R.A., Kentish, S.E., Scales, P.J., Martin, G.J.O. (2013). Quantitative evaluation of the ease of rupture of industrially promising microalages by high pressure homogenization. *Bioresource Technology*, 140, 165–171.

Spolaore, P., Joannis-Cassan, C., Duran, E., Isambert, A. (2006). Commercial applications of microalgae. *Journal of Biosciences and Bioengineering*, 101, 87–96.

Stanley, J.G. and Jones, J.B. (1976). Feeding algae to fish. *Aquaculture*, 7, 219–223.

Stepan, D.J., Shockey, R.E., Moe, T.A., Dorn, R. (2002). Carbon dioxide sequestering using microalgal systems. Final report, US Department of Energy, Pittsburg.

Sueoka, N. (1960). Mitotic replication of deoxyribonucleic acid in *Chlamydomonas reinhardi*. *Genetics*, 49, 83–91.

Sun, L., Wang, C., Shi, Q., Ma, C. (2009). Preparation of different molecular weight polysacchardides from *Porphyridium cruentum* and their antioxidant activities. *International Journal of Biological Macromolecules*, 45, 42–47.

Torzillo, G., Pushparaj, B., Bocci, F.B., Balloni, W., Materassi, R., Florezano, G. (1986). Production of *Spirulina* biomass in closed photobioreactors. *Biomass*, 11, 61–74.

Tredici, M.R., Carlozzi, P., Chini Zitelli, G., Matersassi, R. (1991). À vertical alveolar panel for outdoor mass cultivation of microalgae and cyanobacteria. *Bioresource Technology*, 38, 153–159.

Turpin, V. (1999). Étude des événements physicochimiques et biologiques présidant à la prolifération d'*Haslea ostrearia* (Simonsen) dans les claires ostréicoles de la région de Marennes-Oléron : implications dans la maîtrise du verdissement. University thesis, Nantes.

Unterlander, N., Champagne, P., Plaxton, W.C. (2017). Lyophilization pretreatment facilitates extraction of soluble proteins and active enzymes from the oil-accumulating microalga *Chlorella vulgaris*. *Algal Research*, 25, 439–444.

Venkataraman, L.V., Somasekaran, T., Becker, E.W. (1994). Replacement value of blue-green alga (*Spirulina platensis*) for fishmeal and a vitamin-mineral premix for broiler chicks. *British Poultry Science*, 35, 373–381.

Viso, A.C. and Marty, J.C. (1993). Fatty acids from 28 marine microalgae. *Phytochemistry*, 34, 1521–1533.

Vonshak, A. and Guy, R. (1992). Photoadaptation, photoinhibition and productivity in the blue-green alga *Spirulina platensis* grown outdoors. *Plant Cell Environment*, 13, 613–616.

Waldenstedt, L., Inborr, J., Hansson, I., Elwinger, K. (2003). Effects of astaxanthin-rich algal meal (*Haematococcus pluvalis*) on growth performance, caecal campylobacter and clostridial counts and tissue astaxanthin concentration of broiler chickens. *Animal Feed Science and Technology*, 108, 119–132.

Wang, M., Yuan, W., Jiang, X., Jing, Y., Wang, Z. (2014a). Disruption of microalgal cells using high-frequency focused ultrasound. *Bioresource Technology*, 153, 315–321.

Wang, X.W., Liang, J.R., Luo, C.S., Chen, C.P., Gao, Y.H. (2014b). Biomass, total lipid production, and fatty acid composition of the marine diatom *Chaetoceros muelleri* in response to different CO_2 levels. *Bioresource Technology*, 161, 124–130.

Wolff, A. (2012). L'utilisation des microalgues pour la fabrication de biocarburants : analyse de la chaîne de valeur – contexte français et international. Internship report, Mineure en environnement CERES-ERTI, École normale supérieure Paris, 1–56.

Xie, Y., Jin, Y., Zeng, X., Chen, J., Lu, Y., Jing, K. (2015). Fed-batch strategy for enhancing cell growth and C-phycocyanin production of *Arthrospira (Spirulina)* platensis under phototropic cultivation. *Bioresource Technology*, 180, 281–287.

Zeinali, F., Homaei, A., Kamrani, E. (2015). Source of marine superoxide dismustases: Characteristics and applications. *International Journal of Biological Macromolecules*, 79, 627–637.

Zhang, J., Liu, L., Ren, Y., Chen, F. (2019). Characterization of exopolysaccharides produced by microalgae with antitumor activity on human colon cancer cells. *International Journal of Biological Macromolecules*, 128, 761–767.

Zmora, O., Grosse, D.J., Zou, N., Samocha, T.M. (2013). Microalga for aquaculture: Practical implications. In *Handbook of Microalgal Culture: Applied Phycology and Biotechnology*, Richmond, A., Hu, Q. (eds). John Wiley & Sons/Blackwell Publishing, Hoboken.

Zuorro, A., Maffei, G., Lavecchia, R. (2016). Optimization of enzyme-assisted lipid extraction from *Nannochloropsis*. *Journal of Taiwan Institute of Chemical Engineers*, 67, 106–114.

Index

β-carotene, 52

A

abalones, 43
acid
  docosahexaeonic, 52
  eicosapentaeonic, 91
  okadaic, 101
Aeromonas hydrophila, 50
aerosolization, 18
Alexandrium, 8, 102
algae
  blue, 10
  Klamath, 61
Alveolobiontes, 10
Ankistrodesmus falcatus, 130
Aphanizomenon, 21, 22, 40, 41, 60, 75
Arthrospira, 1, 11, 12, 16, 18, 21, 25, 28, 35, 39–41, 46, 49, 54, 56–58, 60–65, 67, 70, 72, 75, 79, 86, 87, 107, 137
  maxima, 72
  platensis, 11
axenic, 31

B

B-phycoerythrin, 86
Bacillariophyceae, 5
ball milling, 106
Bangiophyceae, 9
batch, 23, 27, 30, 31, 34
bicont, 10
biodiesel, 95, 98–100, 126
biofuels, 94
biogas, 94
biorefinery, 126
bivalve mollusks, 43
Botryococcus braunii, 96, 99

C

canthaxanthin, 52
catalase, 87, 88, 90
Celluclast, 120
Chaetoceros calcitrans, 44–47
Chlamydomonas, 1, 2, 4, 16, 18, 37, 38, 118, 120, 130, 132–135
  reinhardtii, 37
Chlorella, 2, 4, 16, 18, 24, 27, 28, 40, 41, 47, 48, 50, 51, 53–58, 61–65, 67, 68, 71, 73–75, 81, 82, 88, 89, 92, 98, 99, 105, 109–111, 114, 116, 117, 119, 120, 123, 129, 133, 134, 137
  pyrenoidosa, 81
  sorokiniana, 129

Chloromeson, 8
Chlorophyceae, 4
Chlorophyta, 3
Chromophyta, 3
Chrysophyceae, 7
Chrysophycophyta, 5
Cocconeis, 15, 46
 duplex, 46
 scutellum, 15
Crassostrea, 30, 31, 43
 gigas, 30
 plicatula, 43
Cyandophycophyta, 11
cyanobacteria, 1, 3, 10, 11, 13, 16, 18, 21, 23, 27, 28, 37, 39, 40, 41, 43, 46, 54, 56, 60, 61, 63, 66–68, 71, 72, 74–77, 79, 83, 85, 87–89, 123, 125, 132, 135, 137, 138
Cyanophyceae, 11
Cyanospira capsulata, 79
cyanovirin-N, 67
Cyanphyta, 11

D

diarrhetic shellfish poisoning, 8
diatoms, 1, 5
 central or centric, 5
 pennate, 5
dihé, 21, 57, 58
dinoflagellates, 8
Dinophyceae, 8
Dinophysis, 8
Dunaliella, 4, 22–25, 38, 41, 44, 46–48, 50, 57, 68, 72, 75, 88, 89, 94, 96, 123, 133

E, F, G

electroporation, 122
endosymbiosis, 8
Enterobacter aerogenes, 126
enzymatic hydrolysis, 118
epivalve, 5
EPS, 77–81
exopolysaccharides, 9, 68, 69, 71, 72, 77–83
fa cai, 58
Florideophyceae, 9
frustule, 1, 5, 14
green line, 10
 water, 48

H, I, L

Haematococcus, 25, 38, 41, 48, 52, 57, 61–63, 65, 71, 73, 133, 134
Hantzschia, 5
Haptophyta, 10
Haslea ostrearia, 5, 7, 30–32, 65, 90, 137
high-pressure extraction, 114
Homarus americanus, 48
hypovalve, 5, 6
Isochrysis, 34, 44, 47, 48, 50, 57, 63, 65, 89
L112L peptide, 87
Lina Blue, 86
lyophilization, 116

M, N

marennine, 64, 90
 external, 90
 internal, 90
Melosira varians, 16
microalgae, 1, 87, 125
microphytobenthos, 13, 14
microwaves, 113
mode, 23, 26, 28, 31, 34, 40, 46, 128–130
 heterotrophic, 128
 mixotrophic, 128
 phototrophic, 128
Nannochloropsis gaditana, 52, 113, 118

Navicula, 18, 46, 67, 133
directa, 67
Nitzschia, 5, 6, 15, 18, 46, 58, 94
Nostoc, 12, 18, 21, 56, 58, 67, 78, 79
pruniforme, 22

O, P

Oncorhynchus mykiss, 55
Oreochromis niloticus, 50
Parietochloris incisa, 92, 94
Penaeus, 46, 47
japonicus, 46
monodon, 46
subtilis, 46
periphyton, 13
peroxidase, 87, 90
Phaeodactylum tricornutum, 14, 25, 38, 133, 134
Pheophycophyta, 5
photic zone, 14
photobioreactors
ALP, 35
planar alveolar, 38
tubular, 34
physiological forcing, 127
phytoplankton, 12
plankton bread, 58
Porphyridium cruentum, 9, 16, 35, 36, 38, 57, 67, 68, 70, 72, 80, 86, 88, 92, 94, 99, 106, 114, 115, 133
potamoplankton, 13
Prasinophyceae, 45
Prochlorococcus marinus, 12, 14
Pyrrophycophyta, 8

R, S

raceway, 25–28, 34, 40
recombinant proteins, 132
Rhodophyta, 3, 8
shrimp, 43
Skeletonema costatum, 25, 32, 34, 44, 47
Snailase, 120
SOD, 83, 87, 88, 90, 130
solmon, 59, 60
Spirulina, 1
Spirulina sp., 18, 56, 74
supercritical CO_2, 109
system culture
closed-, 21
open-, 21

T, U

tecuitlatl, 57
Tetraselmis, 34, 44–47, 50, 114
suecica, 45
Tilapia, 50
aurea, 50
transgene, 132
Trebouxiophyceae, 4
tychoplankton, 14
ultrasonication, 106, 107, 109, 113, 116
Ulvophyceae, 4
unikont, 10

V, X, Z

Vaucheria, 8, 18
violaxanthin, 89
Volvox, 4, 16
xanthophylls, 52
Xantophyceae, 8
zeaxanthin, 52

Other titles from

in

Ecological Science

2021

JOLY Fernand, BOURRIÉ Guilhem
Mankind and Deserts 2: Water and Salts
Mankind and Deserts 3: Wind in Deserts and Civilizations

2020

JOLY Fernand, BOURRIÉ Guilhem
Mankind and Deserts 1: Deserts, Aridity, Exploration and Conquests

BRUSLÉ Jacques, QUIGNARD Jean-Pierre
Fish Behavior 1: Eco-ethology
Fish Behavior 2: Ethophysiology

LE FLOCH Stéphane
Remote Detection and Maritime Pollution: Chemical Spill Studies

MIGON Christophe, NIVAL Paul, SCIANDRA Antoine
The Mediterranean Sea in the Era of Global Change 1: 30 Years of Multidisciplinary Study of the Ligurian Sea
The Mediterranean Sea in the Era of Global Change 2: 30 Years of Multidisciplinary Study of the Ligurian Sea

ROSSIGNOL Jean-Yves
Climatic Impact of Activities: Methodological Guide for Analysis and Action

2019

AMIARD Jean-Claude
Industrial and Medical Nuclear Accidents: Environmental, Ecological, Health and Socio-economic Consequences
(Radioactive Risk SET – Volume 2)
Nuclear Accidents: Prevention and Management of an Accidental Crisis
(Radioactive Risk SET – Volume 3)

BOULEAU Gabrielle
Politicization of Ecological Issues: From Environmental Forms to Environmental Motives

DAVID Valérie
Statistics in Environmental Sciences

GIRAULT Yves
UNESCO Global Geoparks: Tension Between Territorial Development and Heritage Enhancement

KARA Mohamed Hichem, QUIGNARD Jean-Pierre
Fishes in Lagoons and Estuaries in the Mediterranean 2: Sedentary Fish
Fishes in Lagoons and Estuaries in the Mediterranean 3A: Migratory Fish
Fishes in Lagoons and Estuaries in the Mediterranean 3B: Migratory Fish

OUVRARD Benjamin, STENGER Anne
Incentives and Environmental Policies: From Theory to Empirical Novelties

2018

AMIARD Jean-Claude
Military Nuclear Accidents: Environmental, Ecological, Health and Socio-economic Consequences
(Radioactive Risk SET – Volume 1)

FLIPO Fabrice
The Coming Authoritarian Ecology

GUILLOUX Bleuenn
Marine Genetic Resources, R&D and the Law 1: Complex Objects of Use

KARA Mohamed Hichem, QUIGNARD Jean-Pierre
Fishes in Lagoons and Estuaries in the Mediterranean 1: Diversity, Bioecology and Exploitation

2016

BAGNÈRES Anne-Geneviève, HOSSAERT-MCKEY Martine
Chemical Ecology

2014

DE LARMINAT Philippe
Climate Change: Identification and Projections

Printed and bound by CPI Group (UK) Ltd, Croydon, CR0 4YY